The Ethics and Biosecurity Toolkit for Scientists

The Ethics and
Biosecurity Toolkit
for Scientists

Judi Sture

World Scientific

NEW JERSEY · LONDON · SINGAPORE · BEIJING · SHANGHAI · HONG KONG · TAIPEI · CHENNAI · TOKYO

Published by

World Scientific Publishing Europe Ltd.

57 Shelton Street, Covent Garden, London WC2H 9HE

Head office: 5 Toh Tuck Link, Singapore 596224

USA office: 27 Warren Street, Suite 401-402, Hackensack, NJ 07601

Library of Congress Cataloging-in-Publication Data
Names: Sture, Judi.
Title: The ethics and biosecurity toolkit for scientists / Judi Sture.
Description: New Jersey : World Scientific, 2016.
Identifiers: LCCN 2016015374| ISBN 9781786340917 (hc : alk. paper) |
    ISBN 9781786340924 (pbk : alk. paper)
Subjects: LCSH: Bioethics. | Biosecurity.
Classification: LCC QH332 .S82 2016 | DDC 174.2--dc23
LC record available at https://lccn.loc.gov/2016015374

British Library Cataloguing-in-Publication Data
A catalogue record for this book is available from the British Library.

Desk Editors: Suraj Kumar/Mary Simpson

Typeset by Stallion Press
Email: enquiries@stallionpress.com

Printed in Singapore

Preface

Close to the National Academy of Sciences (NAS) in Washington DC, amongst the trees, sits the Albert Einstein Memorial. It features a large bronze statue of the great man, sitting informally on three curved white granite steps. Engraved on the memorial are some of his famous quotes, including: "The right to search for truth implies also

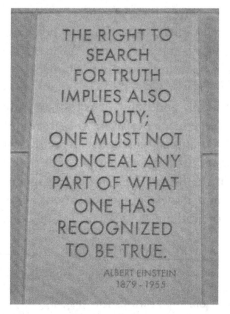

Photo: J. Sture

a duty; one must not conceal any part of what one has recognized to be true." This particular quote is even repeated on the wall of the NAS at its headquarters at 2101 Constitution Avenue NW just around the corner.

Why then, do the NAS and other prominent institutions and bodies, now endorse work that supports and promotes potential restrictions in scientific practice and communications? Is this not an attack on academic and scientific freedoms? Why is such work being funded by several western governments? Is it not our right, as scientists, to go wherever science takes us, and to publish whatever we find? What interests could possibly undermine these rights?

Anyone who follows the scientific press, not to say the general news media, cannot have failed to see recent challenges to academic and scientific freedoms as a result of the H5N1 situation in 2011 and 2012.[1] Those involved in maintaining and strengthening international security, along with those engaged in the prevention of terrorism, believe that the current drive to increase the reach of biosecurity into the daily lives of more scientists is clearly warranted. Much of the general media seems to share these concerns. The question we must now answer is this: do they have a point?

I assume that most scientists reading this book are appalled by this state of affairs. So why is this happening? What can scientists do about it? *Should* scientists be doing something about it? If so, *what* should they be doing?

My answer is simple and blunt. It is probably unwelcome. But it is this: scientists *must* engage positively *and proactively* with the needs of national and international biosecurity. If we do not, then someone else will do it for us. If that happens, we will only face further and worse problems in the future. Part of our engagement process must include an open and honest discussion about our 'rights' and the nature and scope of academic and scientific 'freedoms'.

---

[1] See *Nature* Special Issue 12 January 2012. Available on: http://www.nature.com/news/specials/mutantflu/index.html. Accessed 21/12/15; and *Science* Special Issue 22 June 2012. Available on: http://www.sciencemag.org/site/special/h5n1/index.xhtml. Accessed 21/12/15.

Put simply, the world is changing, for scientists as well as for everyone else. The science community can no longer continue under the blanket assumption that all scientific work is carried out for the good of humankind simply because 'scientists know best'. The rise of a public awareness of human rights, the readiness of the public and governments to engage in litigation in pursuit of their rights and requirements, a 'compensation culture' mentality, the increasing recognition of ethics as a key aspect of public life, along with the increasing scope of social responsibility at corporate and personal levels, *are with us to stay*. All of these have an impact on science and scientific practice. This impact is already felt,[2] and it is only going to escalate.

Whether we like it or not, scientists are increasingly confronted with questions about their work and its relationship to the social good. The H5N1 publication controversy is a good example of the kind of interrogative interventions that the public, science stakeholders and governance bodies are beginning to make in science policy and practice. Science is, arguably, being held to account like never before. We need to respond to this effectively. Putting our heads in the sand and shouting about our strengths while never acknowledging our weaknesses is not enough. We can no longer avoid explaining ourselves to others in detail and repeatedly. We can no longer rely on our position or our expertise to avoid public accountability at a level we have probably not previously encountered.

As well as responding to increasing security issues themselves, today's scientists are also faced with the implications of global communications technologies and the way these open science up to the public. The effects of the rise of social media, the speed with which ill-informed 'information' and 'fact' can spread around the world like some sort of (ironic) contagion, cannot be lost on readers. The ever-increasing reach of social and news media, platforms that give the public a voice and the ability to spread opinion in ways that were not previously possible, is something that increases scrutiny of our work.

---

[2] World Health Organisation (2012) *Report on technical consultation on H5N1 research issues* Geneva, 16–17 February 2012. Available on: http://www.who.int/influenza/human_animal_interface/mtg_report_h5n1.pdf. Accessed 21/12/15.

It also increases scrutiny of *we ourselves* and our motives. We cannot avoid this.

Yes, scientists usually have a greater and deeper understanding of 'the issues' around science and research than does Joe Public. But in our media-led environment, like it or not, it is a fact to be regretted that today *opinion* is often judged to be on a par with, or even more important than, *expertise*. In fact, the role of authority and expert knowledge is being challenged in all sorts of sectors and contexts. Science is not the only monolithic 'body' to be subject to this sort of pressure. The public, informed or not, is now in a position to challenge science in ways that have not been possible in previous generations. We already have a trained 'army' of ambulance-chasing personal injury lawyers in the public sphere. How long will it be before we have an army of scientist-chasing lawyers, with an eye on the ever-increasing compensation market? In today's world, human rights seem to trump responsibilities at every turn and there is always someone on hand to convince some individual of his status as the victim of someone else's wrongdoing. Unless we as scientists take action to protect our work from unreasonable accusations and pressures, we will lose the argument. If this happens, cries of 'where there's blame, there's a claim' will soon be echoing around laboratories just as they do around Health and Safety offices and hospital emergency rooms.

This book does not include a lengthy debate about the rights and wrongs of this state of affairs, although some arguments are addressed briefly. Instead, it is written in the context of this situation *being real*, even if unrecognised and unwanted. In response, it presents a set of ethics-based questions — The Ethics Toolkit — that, when applied to scientific work in the search for any potential for dual use,[3] will allow consideration of the science in relation to social responsibility and the public. When this 'social responsibility ethics' approach is combined with the search for dual use potential in scientific work (inherent in the Toolkit questions), the result is a biosecurity framework that can equip scientists to answer and to adapt to this state of affairs in order to carry on working as freely as possible in an accountable manner.

---

[3] 'Dual use' refers to the potential for any scientific work to be applied to hostile as well as benign purposes.

Readers may ask, why do we not fight this situation? That's a good question. However, to this I would say, we cannot fight it — we are already living in it. Security concerns are here to stay. We cannot live in a world which embraces certain perspectives on human rights and ethical behaviour and at the same time make our own activities exempt from the consequences of this. Crucially, if we as scientists do not take a lead in this situation, *someone else will do it for us.* Unfortunately, the others doing it for us are likely to be politicians, 'community leaders' of various sorts, special interest groups, competing commercial lobbyists, a vocal minority group or groups, or worse, a set of committees comprising all of the above.

So how do we avoid this undesirable situation? I would say that if we as scientists can apply a practical ethics-based biosecurity framework to our work, and *be seen to do so* on a regular basis as a routine part of our practice, then we are more than half way to meeting the concerns of non-scientists (including many in the security community) about the security aspects of our work. By showing how we identify and respond to biosecurity issues in our work just as routinely as we already do to biosafety concerns, we will stand a chance of remaining in *the* position of influence over our own work.

To be able to do this effectively, we need to incorporate suitable and appropriate ethics training into our education as scientists, from school up. Both biosafety and biosecurity are ethically-mediated concepts. I will show in the following chapters how biosafety can be seen as an ethically-mediated framework and demonstrate how biosecurity can be taught, understood and implemented in the same way. This ethics education needs to be continued and developed post-qualification through Continuing Professional Development (CPD) classes and in best-practice sharing sessions between peers and institutions. Importantly, this means that scientists can be trained to be ethically aware and ethically responsive throughout their careers.

Embedding ethics education into courses and CPD programmes also builds an ethics capacity within the discipline. Students and scientists will be able to pass on their knowledge and ethical practice to their juniors in turn. This is not difficult. Before you started work in the lab, you had to be trained in biosafety — a system or framework of *containment.* Now it is embedded in your approach to daily work.

You can usually recognise a biosafety issue and respond effectively to it. If you are already 'doing' biosafety, you are already 'doing' ethics. Ethics is that easy.

The ethics of biosafety involves, for example, issues of consent, autonomy, privacy, doing good, doing no harm, and responsible professional behaviour. Now you just need to be shown a 'containment framework' with which to apply ethics as a means to achieve biosecurity. If you do the ethics effectively, you will have operational (even though not perfect) biosecurity. It's that simple. In reality, the practical application of a biosecurity ethics framework is not any different in terms of limitations on work to the *biosafety* risk management framework that already exists. All this book provides is an *added strand* of risk management — that of applied biosecurity — as ethics, that's all.

As scientists, we do not question or worry about the limitations imposed on us by the daily accommodation of biosafety processes. This is because we understand them and can see how they help us. Neither should we worry about similar limitations put on us by *biosecurity* processes. Biosecurity protects us in the same way that biosafety does. It just goes 'beyond the laboratory door.' That last point is crucial. Accepting that biosecurity extends outside the laboratory through people, their minds, written materials and physical materials, is key to understanding the nature and scope of biosecurity risk. Both biosafety and biosecurity concepts are built around issues of containment to avoid harm, and both concepts require scientists to compromise their personal rights and autonomy in the pursuit of safety of some sort at some level.

By adopting an ethics-based biosecurity approach alongside our existing *biosafety* approach, we can be confident that we are addressing our ethical and social responsibilities in an accountable way. If we are challenged by non-scientists, we can show what practical steps we have taken to meet the identified biosecurity risks, just as we do with biosafety risks when questioned. Others can then judge us against these standards which we have voluntarily adopted (or been legally required to do). Sooner rather than later, biosecurity practice will be regulated by law — so scientists need to be in the lead in promoting biosecurity frameworks before the lawyers, politicians or the vocal, one-issue or uninformed lobbyists get hold of them.

This book presents a biosecurity framework that is operationalised through ethical principles. This is a sound, informed basis for codification into law when that time comes. Along with the framework of ethics, this book provides, at various points, practical 'Intervention Point' questions that can be posed and answered in an ethically effective way. I do not seek to provide answers to every conceivable biosecurity situation in which scientists may find themselves, but I do offer a set of principles — operationalised through questions — to assist scientists in their day to day work that will enable them to recognise and respond effectively to biosecurity issues.

Adoption of such approaches cannot, of course, remove all risk of unwanted outcomes. It can, however, support us in reasonably foreseeing, at the time of the work, *some* potential unwanted outcomes. It also enables us to plan and take action to mitigate the risks of those to the best of our ability, at the time of the work. In addition, when things go wrong — which they always will, as we are human — we can review our ethics-based policies and processes and see what areas need to be revised or changed to avoid repeats in the future, *by using the same ethics framework*.

Still not convinced of the need for all of this? If you remain to be persuaded, may I suggest that you answer all of the following questions as honestly as you can? If you would like to see some 'real life' evidence of the sort of issues that scientists have faced or created already, then please also go to Chapter 8 where I have provided some real life cases of biosecurity problems.

Meanwhile, here are some straight-to-the point dual use related questions:

- Are you indemnified against the outcomes of current or future unintended or intentional misuse of your work?
- Is your institution insured against claims arising from the unintended or intentional misuse of your work?
- Do you know that if you work in any aspect of chemistry and/or biotechnology, even in schools, your work is subject to the terms

of the Chemical Weapons Convention (CWC) and the Biological and Toxin Weapons Convention (BTWC)? (Yes, it really is).

- Do you know and understand your personal and professional responsibilities under the BTWC and the CWC?
- Do you know and understand your liabilities under these conventions, and under the national law of your own country in respect of your work?
- Do you believe that your work poses no biosecurity threat to others or to the environment?
- How do you know that?
- At what points in time (from starting to plan the work) and space (geographically) do you believe that your responsibility for any unintended or intentional outcomes of your work ends? Try drawing a diagram to illustrate your answer. (Hint — 'the end of the grant' or 'once it's safely published', or 'when I've won the Nobel Prize' are not good enough responses).
- Do you know and understand the biosafety risks inherent in your work? (I hope you answer 'yes' to this).
- Do you know and understand your biosafety responsibilities? (I assume your answer to this is 'yes').
- How do you know this?
- What frameworks do you use to recognise and understand these responsibilities? (I assume you can answer this easily).
- Do you know and understand the *biosecurity* risks inherent in your work?
- Do you know and understand your biosecurity responsibilities?
- How do you know this?
- What frameworks do you use to recognise and understand these responsibilities?
- Do you know and understand the *ethical* risks inherent in your work?
- Do you know and understand your ethical responsibilities?
- How do you know this?
- What frameworks do you use to recognise and understand your ethical responsibilities?

- Finally, if you were to be held *personally*, *legally*, *morally*, and *financially* responsible, for *all* outcomes of your work in the future (including misuse of your work), *would you still be doing it tomorrow?*
- If *you* are not responsible for the outcomes of your work *then who is?*

Extra content for this book can be found on the following page under the 'supplementary material' tab: http://www.worldscientific. com/worldscibooks/10.1142/q0027#t=suppl

About the Author

Judi Sture is a biosecurity education consultant and biological anthropologist. Her first career was as a State Registered Nurse (Leeds General Infirmary, UK). She gained a First in her BSc (Hons) degree in Archaeology in 1997 at the University of Bradford's Department of Archaeological Sciences, where she began to specialise in the study of human remains. Having won a PhD studentship in the British Academy/AHRB's national open competition in 1997, she completed her PhD in biological anthropology at the University of Durham in 2001. Her doctoral thesis focussed on associations between the environment and human birth defects. Her work in biological anthropology triggered an interest in research ethics, which she developed when she took up a position as Lecturer in Research Methods at the University of Bradford in 2001. She developed a range of postgraduate courses in Ethics which have been taken by several thousands of scientists, social scientists, and a range of staff and students from other disciplines since then.

Judi was appointed Senior Lecturer and Head of the Graduate School at Bradford in 2006, developing, leading, and overseeing cross-institutional postgraduate research training programmes, including in Ethics, at the University until 2015. In 2008, due to her ethics expertise, Judi was invited to be a Research Fellow in the University's Bradford Disarmament Research Centre. Following this appointment, the centre won major funding from the Wellcome Trust for a project promoting dual use biosecurity in the life sciences: *Building A Sustainable Capacity In Dual-Use Bioethics*. This five year project was undertaken with colleagues from the Bradford Disarmament Research Centre, the Universities of Bath and Exeter and Australian National University, Canberra.

Judi has since gained personal funding from the Economic and Social Research Council, the Japan Society for the Promotion of Science, the UK Ministry of Defence (Defence Science and Technology Laboratory), the US Department of State, the US Department of Defense and the Public Health Agency Canada (PHAC). This work has involved, among other outputs, the development of biosecurity courses for biosafety professionals, scientists and science students and the development of ready-to-deliver country-specific biosecurity training courses (*The National Series*) for a range of countries including Algeria, Armenia, Azerbaijan, Egypt, Georgia, Jordan, Kazakhstan, Kyrgyzstan, Libya, Morocco, Pakistan, Saudi Arabia, Tajikistan, Tunisia, and Ukraine. Judi's work in promoting biosecurity through ethics has taken her around the world including work in Australia, Indonesia, Tunisia, the USA, Canada, Japan, Switzerland, the Netherlands, Denmark, Italy, Spain, Georgia, Ukraine, and others. She has both chaired and presented her work at many side events at the United Nations in Geneva during meetings of the Biological and Toxin Weapons Convention and has made formal statements to the assembly there. In July 2012 she was invited to formally address the assembly of the BTWC in Geneva on biosecurity education. In September 2014 she was invited to speak on the same subject at a major meeting at the Organisation for the Prohibition of Chemical Weapons (OPCW) in The Hague.

Judi has organized and led a number of international expert meetings at the University of Bradford between 2010 and 2013, and has edited the *Yearbook of Biosecurity Education* arising from these meetings. She has worked with colleagues from the US National Academy of Sciences and the American Association for the Advancement of Science in Washington DC. She has also been a member of the Advisory Board for projects led by Georgetown University developing healthcare programmes in Yemen. Since 2012 she has carried out consultancy work for the US Department of State and Department of Defense. As part of this she has undertaken training for herself at Sandia National Laboratories in New Mexico. She has planned and led two biosecurity education workshops for scientists in Baghdad, Iraq and has written a policy guidance paper for the Department of Defense advising on new ways of effective collaboration between US and Middle Eastern and North African scientists. Judi was also part of a small Working Group that revised a policy statement on *Managing Risks of Misuse Associated with Grant Funding Activities* on behalf of the UK's Biotechnology and Biological Sciences Research Council (BBSRC), Medical Research Council (MRC) and the Wellcome Trust in 2013. She is currently working on further international educational activities for the US Department of State.

Contents

Chapter 1

Why Do We Need Biosecurity Ethics in Science?

Why Do You Need to Read this Book?

Are you aware that if you work in any aspect of chemistry or biology, your work is subject to the terms of the Chemical Weapons Convention (CWC) and the Biological and Toxin Weapons Convention (BTWC)? Amazing at it seems, your work is actually covered by these international treaties even if you are 'just' a science teacher in school, a science lecturer at college or university, someone who works alongside scientists in supporting and facilitating their work, or if you are involved in the devising of science policy and practice. These two treaties apply to you if you engage in any scientific work in biology or chemistry. At any level.

Do you know and understand your personal and professional *responsibilities* under the BTWC and the CWC? Do you know and understand your *liabilities* under these conventions, and under the national law of your own country in respect of your work? Do you believe that your work poses no biosecurity threat to others or to the environment? How do you know that? Can you be absolutely sure, or even reasonably sure, that the work you are doing or facilitating is *not* being carried out for non-peaceful or harmful purposes, or that it will not be repeated in the future for hostile purposes? Can you be

1

absolutely sure, or even reasonably sure, that the people with whom you work, or who you teach, are intent on carrying out scientific work only for peaceful purposes? What responsibilities do these scenarios place on you as a scientist, manager, tutor, Principal Investigator (PI), grant-holder, colleague or other beneficiary of funding? What do you think about the recent (2011 onwards) debates in the science and mainstream media about the risks of 'dangerous' biotechnology?[1]

I know that these questions seem alien to most scientists. I recognise that the overwhelming majority of scientists have entered the profession in order to work to the benefit of human kind. However, I have also worked with some scientists, representing a minute proportion of the global scientific community, who were obliged to work on projects in the past that were not being developed for peaceful purposes. Their work had the potential to cause massive human suffering. It only takes the work of a relatively small number of individuals to cause a disproportionate amount of harm. In some cases, the former weapons scientists with whom I have worked did not know or understand the end goals to which their work was contributing. Under certain types of government, it is not always clear what the end-goals of scientific work are, and it is perfectly possible to be working on hostilely-intended projects and not be aware of it. Either way, these scientists were engaged in illegal scientific activities, working on biochemical weapons — whether these were used or not. More common, of course, is the likelihood that some ill-intentioned person, group or government will engage in the *future* misuse of *benignly intended* science.

We all need to think about this today. The 'bad stuff' is not all in the past. It only takes a regular glance at the evening news and the internet to see the threats that face us today from increasingly sophisticated enemies. It may not be too long before our tradition of open education for students from around the world is challenged in the name of national and international security. We need to take steps

---

[1] For example, the so-called H5N1 Debate, see *Nature* Special Issue 12 January 2012. Available on: http://www.nature.com/news/specials/mutantflu/index.html. Accessed 21/12/15; and *Science* Special Issue 22 June 2012. Available on: http://www.sciencemag.org/site/special/h5n1/index.xhtml. Accessed 21/12/15.

now to show how we are recognising and addressing security issues in science — and the use of an applied ethics framework can assist us in doing this.

One of the principal prohibitions of both the BTWC and the CWC is the assistance, encouragement or inducement of anyone to engage in the development and use of biological and chemical weapons.[2] Think about this. If we pass on or share the means of using science for hostile purposes to our colleagues and students, where does the responsibility for misuse lie and end? We work in a global education and commercial environment — we need to think about how we share our work, while doing our best to maintain open scholarship. We need to think about how we recruit students and staff, while trying to maintain fair practices and meet our legal responsibilities under national law. None of this is easy, but in today's world of terrorism, unstable states (the biggest risk), identity politics, and instant global communications, we need to think about how these impact on our long-treasured scientific freedoms and academic 'rights'. I would argue that we need to adapt to these circumstances and that we cannot simply continue to work 'as usual' without taking mitigating actions to identify and reduce biosecurity risks. Of course, in order to mitigate risk, we have to recognise it in the first place.

Scientists have a unique role as the 'guardians of science.'[3] The current security-focused approach to 'dangerous' science has thrown up all sorts of problems for scientists and has not necessarily resulted in reduced risk in the eyes of the security community. Plus, if we need to know how to make biological and chemical weapons in order to counter them effectively, how does 'reducing risk' fit into that picture?

---

[2] BTWC Articles III and IV. Available on: http://www.opbw.org/convention/conv.html. Accessed 21/12/15; CWC Article I. Available on: https://www.opcw.org/chemical-weapons-convention/download-the-cwc/. Accessed 21/12/15.

[3] McLeish, C. and Nightingale, P. (2007) Biosecurity, bioterrorism and the governance of science: The increasing convergence of science and security policy. *Research Policy* 36, 1635–1654. Available on: http://www.sussex.ac.uk/Units/spru/hsp/Draft%20Convention%20supporting%20docs/HSP%20papers/CMcL_RP_2007.PDF. Accessed 21/12/15.

One of the key issues, often missed it seems, is that we need to look at *motivation* and *end-purpose* when we consider the security risks in science. As much of this is culturally-determined, we also need to recognise that not everyone around the world shares our views on what does and does not constitute 'dangerous' science or even 'risk'. If the security community and much of wider society continue to insist that certain aspects of science are 'dangerous', then we will simply dig a deeper and wider gulf between science and the rest of society. It is surely better for both science and governance professionals to engage together in a joint bid to protect science from misuse by rogue individuals, rogue groups, and rogue states. As I mentioned above, we only need to watch the evening news to get an idea of who may fit into these categories at any one time.

Ethics as a Security Tool

Like any professional workman, you need a good toolkit. 'A workman is only as good as his tools', and this goes for ethics as well as for research and work capability. The Ethics Toolkit — a set of questions that applies ethical principles to your work — will help you to identify ethical problems, make ethically sound decisions about your plans and actions, and will provide you with a framework around which you can design any work or research project that will reflect a principled sensitivity to the rights of others.

Taking an applied ethics perspective does not mean *stopping* or preventing work; The Ethics Toolkit simply offers a way of assessing scientific activity for motive (social, political, economic, technological, and so on), possible misuse opportunities and overall *potential for* or *risk of* misuse. Benignly-intended work should be allowed to continue. 'Difficult' work should continue wherever possible, but we need to be big enough to consider that some of our traditional routes of publication and sharing may need to be modified occasionally. Rogue scientific activity should not be enabled or facilitated either directly, or by accident, or by failure to recognise risk. We need mechanisms that enable us to identify potential or actual rogue activity at the earliest possible stages. If we can recognise the *potential* for misuse of our work our-selves, this can forewarn us of the possibilities of *actual* misuse.

Bearing all this in mind, it seems that scientists can no longer avoid public concerns and claim the rights of scientific and academic freedom, without at the same time showing how they recognise and mitigate risk in their own work and that of their colleagues. While this is arguably an outrageous situation in the eyes of most of us, it is simply the direction in which society is moving. In order to mitigate the challenges raised by such increasing accountability, an applied eth-ics approach can be employed. Once applied and maintained as an ongoing activity, the security community, governments and the public will be able to *see* how actions are being taken to protect science by scientists. They may not *agree* with all the ethics-based decisions, but at least we will have a practical applied ethics framework that can sup-plement existing precautionary measures and controls, allowing us to *show* how we are 'being ethical'.

I do not pretend to have all the answers, but I suggest that by employing The Ethics Toolkit, which can be implemented alongside other decision-making processes, we will be able to review and assess our practices and our assumptions in such a way as to mitigate some of the risks for misuse that emerge in today's world of science. This is because the Toolkit enables its users to consider their responsibilities alongside the rights of other people, granting equal status to both elements of the equation.

Biosecurity in a changing world

We live today in a world where security concerns are more pressing than ever. Security in science (not to mention science *as* security) is a growing concern not only for the public, but also for governments and scientists. As scientists, we need to raise our awareness of biosecurity as a concept — in the face of the rapidly expanding biotechnology indus-try worldwide, as well as in the face of the increasing convergence of biology and chemistry that goes along with this. We need to learn how to take the necessary steps to protect our work, ourselves, our societies and our world from those who wish to misuse 'our' science.

If you have not already done so, please go back and read the Preface before reading further in this chapter, especially the questions at the end. This will help you to realise that science security may not be as straight forward as you thought, even if you have been practising science for years.

For those of you who are new, or relatively new, to science as a profession, or even as a hobby, let's look at the sort of questions we are being faced with today. Why are scientists 'inventing' new pathogens? How can that be safe? Why is this allowed? How do we know that science is really safe? Is it right to make a pathogen spread more easily? Who do scientists think they are?

Some of these questions simply did not arise for previous generations. Other questions have always been around, but were not considered as 'important' as the outcomes of the science ('the end is worth the means'). In the past, scientists were 'up there' socially, whether as doctors, surgeons, professors, atomic bomb-builders or even astronauts. This lofty social status protected scientists in previous generations, relatively speaking, from many of the questions or concerns that may have been felt by the general public. When science could be seen as something that made life better, who wanted to question it or its practitioners deeply?

Today, things have changed. People are more aware than ever of their rights. They are better educated than their forebears. They are ever more ready to hold 'authority' to account, whether it be for compensation, or for effecting social change, or for some political or commercial gain. We have Twitter, email, blogs, streaming news on our phones, Facebook and all that goes with these 'wonders' of communication and social media. If we make a big mistake, everyone will be talking about it before the end of the day. If someone wishes to share information for hostile purposes, it can be done at the touch of a 'send' button. In the early 21st century, any member of the public can ask a big question of scientists. Before we know it, that question is on the front page and we have to answer it. It does not matter if it is an uninformed or a misinformed question. It is the questioner's 'right' to have an answer. If he does not like the answer, he can come back and ask it *louder*. An audience is engaged, following the unfolding drama. Newspaper comments sections fill with 'debate' and a game of Chinese Whispers ensues, in which the truth, whatever that is, is lost, discarded, ignored, misrepresented or not recognised at all. Scientists come off worst, and 'Manonthestreet@peoplepower' is not satisfied with the answers he received. He is convinced there is some sort of cover-up in place. He is sure that 'they' are doing 'secret things' that they don't want 'us' to know about. He wants something to be stopped. His favourite newspaper runs it on the front page for a few days. At work, the Vice Chancellor, the Dean or the Head of Department start visiting your lab every day. The boss is always on the phone. You have to justify yourself not only at work but also in public. You start to think it might be easier to just run a grocery stall in the market than spend your life searching for a cure for cancer (or whatever your main aim is). Why do you have to go through all of this?

I mentioned in the preface that today, opinion seems to be as important as expertise, if not more so. Unfortunately, that's just where we are. As scientists we have much of the expertise. We are used to being 'in the know' and we often view the public's apparent lack of understanding as something that enhances our own social mystique (even if we do not recognise this). But, does this mean that we can simply dismiss the concerns of others as plain wrong, or as uninformed, or as misinformed opinion? I would say that the answer to that is 'no'. Crucially, we ought to be prepared to challenge *ourselves* when some of these 'uninformed' questions actually raise good points. Are we big enough to do this? Or are we more concerned to maintain our status as members of the only stronghold of expertise that cannot be questioned?

With expertise comes responsibility. Today, part of that responsibility involves explaining and accountability. We may have to make the same explanation over and over again for years. But if that sort of accountability is part of what is necessary in keeping the public and governance bodies 'onside' with our work, let us do it. By taking an applied ethics approach such as that outlined in this book, we will be using a recognised framework to explain and give account of how we incorporate biosecurity in our work. We will also be able to use this framework to evaluate not only our own work but that of others, highlighting existing problems and seeing the potential for future issues that can be mitigated in good time. Science is not just for scientists.

It is in our interests to protect science from those who would seek to misuse it. The days of 'we know best and that's all you need to know' are long gone. We are, in the end, all working for the public one way or another.

What is Dual Use?

The term 'dual use' is a contentious one. Although initially coined to refer to the militarisation of civilian technology, it has acquired a wider meaning, incorporating any 'double use' of science or technology that transfers it from a 'good' purpose to a 'bad' purpose. Many authorities dispute the usefulness of the term today, as it has negative and judgemental implications in the eyes of scientists. However, efforts at finding a blanket term with which to replace it have not reached a consensus. Other terms which are suggested to replace 'dual use' include 'responsible conduct of research' and 'culture of responsibility' amongst others. As is obvious from these suggested replacement terms, they do not actually refer to the problem but rather to proposed mechanisms with which to tackle the problem. This is not the place to debate this, but I will retain the term 'dual use' here as I am speaking about the inherent *potential for misuse* of much science and technology. Of course, I am also talking about mechanisms to avoid dual use activities and to raise awareness of the risks, but rather than argue over terminology I will just cut to ideas about how to tackle the problem.

The simplest example of dual use is the misuse of a kitchen knife. It is designed to cut up food. That is its primary purpose. However, it can also be used to kill another person. This is a potential 'dual use' of a household knife. The *application* is dual use, not the knife itself. In the same way, the application of science to hostile purposes does not say anything negative about the science itself. It is the application of it that is wrong (assuming that we can agree that killing or harming someone is wrong).

We also need to remember that dual use application can occur by accident — a fact that most of us are probably unaware of. Misuse may result from hostile motivation or from negligence leading to an *opportunity* for misuse. For example, a scientist or group of scientists may actively plan to use some scientific work for hostile purposes, or alternatively, a simple failure of lab security or biosafety controls may enable rogues scientists (or others) to acquire pathogens for hostile purposes that were not part of the original work that enabled them to acquire the pathogen. We could say the same thing about the *knowledge* that would enable someone to misuse science in a hostile manner. Who is responsible for this?

A dangerous pathogen is only dangerous when it is uncontained; a scientist working on a dangerous pathogen is actually working on a pathogen that is not necessarily dangerous if all the right protocols are followed (see Chapter 8 for some caveats). The questions here would be — *are* all the right protocols being followed fully and without breach? And *are the current protocols sufficient and effective?* In the same way, we need to look at knowledge, experience and practice as equally important and necessary means to effect a biological or chemical attack for hostile purposes. This is the point at which we hit the 'scientific freedom' buffers.

As scientists, we are best protected from accusations of misuse — or the potential for future misuse — if we can show how we have identified and considered the dual use risks that our work with a dangerous pathogen or process and the knowledge and practice in handling it may raise. If scientists clearly show that their intention is benign, including the steps they have taken to prevent or circumvent dual use, society can be assured that they have recognised the risks of their work being misused for hostile purposes by others — or indeed by themselves or others at a later date.

Scientists are thus the best 'guardians of science' against misuse. It is in our interests as scientists to protect science from misuse. If we do not, then we will attract massive public opprobrium and the subsequent loss of our funding, our work and maybe even our careers, when something goes wrong. It is perhaps not too great a leap of the imagination to also assume that at some point in the future, scientists may be prosecuted for the misuse of their work downstream if they did not take reasonable precautions at the time of the work to avoid future opportunities for misuse. While this may sound ludicrous, we need only consider the group of seismologists who were jailed in Italy in 2012 for manslaughter to see that this is already upon us. The verdict against these scientists was based on how they had assessed and communicated risk before an earthquake that hit the city of L'Aquila on 6 April 2009, killing 309 people.[4] While the unfortunate scientists had this conviction overturned in 2014,[5] who would want to go through this?

A simple definition of dual use potential in the life sciences can be this: 'Dual use potential is the inherent capability in any biological or chemical scientific activity that could enable it to be applied for hostile as well as benign purposes.'[6] There are other good definitions as well, but this one sums it up sufficiently for now. We need to bear in mind as well that the 'old' definition of dual use tended to refer to the *military* use of civilian science — current definitions can include such uses, but also encompass the possibilities of misuse by non-military groups or individuals as well as military dual use.

In 2007, McLeish and Nightingale[7] developed a useful conception of dual use. They clarified the differences between the 'traditional' view of dual use and a revised approach that does not infer 'danger' where there may be none — a welcome approach in the eyes of scientists. They showed how the traditional view of dual use sees it as an issue of *technology transfer* in which science is transferred from a 'good' use to a 'bad' use by hostile persons. By emphasising that science cannot in itself be 'bad', rather it is some of the *uses* or *applications* of science that can be 'bad' — they showed more clearly how technology (of which science is a part) can be viewed not only as

---

[4] Nosengo, N. (2012) Italian court finds seismologists guilty of manslaughter. *Nature* 491, 15–16. DOI:10.1038/491015a. Available on: http://www.nature.com/news/italian-court-finds-seismologists-guilty-of-manslaughter-1.11640.

[5] Abbot, A. and Nosengo, N. (2014) Italian seismologists cleared of manslaughter. *Nature* 515, 171. DOI:10.1038/515171a. Available on: http://www.nature.com/news/italian-seismologists-cleared-of-manslaughter-1.16313.

[6] Dual use can occur in any field, but I am concentrating here on biology and chemistry.

[7] McLeish, C. and Nightingale, P. (2007) Biosecurity, bioterrorism and the governance of science: The increasing convergence of science and security policy. *Research Policy* 36(2007): 1635–1645.

artefacts but also as *knowledge*. Further, they considered how these scientific artefacts and knowledge can only operate in socially-mediated contexts in which benign or hostile motivations can flourish. Crucially, by taking this different view of dual use, as *technological convergence*, controls can be applied that will allow hostile purposes to be identified and prevented from completion, while leaving benign scientific work unimpeded. McLeish and Nightingale used the examples of wildly differing items that are derived from the same technology, such as bicycles and sewing machines or prohibited weapons and vaccines. This is a view that underpins The Ethics Toolkit approach to recognising and addressing risk in biochemical security contexts.

This definition of dual use as *technological convergence* clearly shows how dual use is best conceptualised — science is not a danger, but the people who want to misuse it *are*, as are any opportunities for them to apply misuse in practice. Dual use in science can therefore be defined as the intended or unintended *misuse* of that science, arising from motivation. *Potential for dual use* refers to the opportunity and possibility of peaceful science being misused in the future.

Much benignly-intended science holds the potential for misuse. We are simply not accustomed to looking for this. It can occur by mistake, be facilitated by poor oversight, by lack of knowledge and understanding, or it can be deliberate, either from the start, or developed once a dual use potential has been identified. Whatever the source, the outcomes of hostile dual use are bad for the public. The great achievements of the 19th century scientists such as Koch, Lister, and Pasteur on the properties and characteristics of micro-organisms were not only foundational in the establishment of effective public health measures that saved millions of lives, but also formed the basis of the subsequent development and use of biological weapons. What makes us think that history will not repeat itself with 21st century biotechnology advances? As practitioners of science, we therefore need to be aware of the potential for dual use of our own work and of the work of our colleagues and others. This is a professional and ethical responsibility that we need to pick up as guardians of science. Dual use awareness, or misuse awareness, should be part of every scientist's skill set.

The British, the US, French, South African and other western governments all developed biological/chemical weapons programmes in the 20th century, as well as the Russian, German and Japanese governments.[8] It is very easy to label others as 'acting dangerously' only to find that we are doing the same, or similar things. It is particularly easy to blur the lines if a government wants to develop a defensive capability against biochemical weapons, but says that it will not use its knowledge offensively. We need to know how to develop biochemical countermeasures in order to mitigate the adverse outcomes of them when they are used on us. How does this make you think about 'defensive' programmes as an option? Have you ever considered how your work could be used by state-level agencies in this sort of scenario? If this ever happened to your work, would you be able to stop it? Would you want to stop it? Where would your responsibilities lie if your work was to be developed into a biochemical weapon?

Modern Biotechnology and Dual Use Risks

In 2000, Matthew Meselson of Harvard University's Department of Molecular and Cellular Biology, said:

> Every major technology — metallurgy, explosives, internal combustion, aviation, electronics, nuclear energy has been intensively exploited, not only for peaceful purposes but also for hostile ones. Must this also happen with biotechnology, certain to be a dominant technology of the 21st century?[9]

Meselson went on to comment:

> If this prediction is correct [citing a 1989 prediction of biotechnology as a source of security problems], biotechnology will profoundly alter

---

[8] See Dando, M. (2006) *Bioterror and Biowarfare; A Beginner's Guide,* Oxford Oneworld Publications.
[9] Meselson, M. (2000) Averting the hostile exploitation of biotechnology. *CBW Conventions Bulletin* 48, 16–19.

the nature of weaponry and the context within which it is employed.... During the century ahead, as our ability to modify fundamental life processes continues its rapid advance, we will be able not only to devise additional ways to destroy life but will also become able to manipulate it — including the processes of *cognition, development, reproduction,* and *inheritance....* Therein could lie unprecedented opportunities for violence, coercion, repression, or subjugation.[10] [Italics mine]

While Meselson refers here to biotechnology, his words could equally apply to many other fields of scientific endeavour. So how can we avoid, or reduce, the risk of adverse outcomes arising out of our *good* work? How can we be *sure* that only good outcomes will come out of our work? *Can* we be sure that we can ensure only good outcomes? (Hint — no, we can not, but we do have to *try*).

As scientists, we are uniquely privileged in being able to affect the lives of millions of other people, often in ways that were unimaginable even just a decade ago. Of course, we all want to affect those lives positively, not negatively. In our own eyes, all of our work is positive. Nobody usually enters science to cause harm (which may not always be the case today, sadly). But in our rapidly-advancing fields of research, especially of biotechnology, biochemistry and at the various sites of biological and chemical convergence, we have the capability to develop and implement a lot of work that can do both good *and* harm, depending on how it is developed and applied. This has always been the case, but today our skills and research potential are such that even greater risks of mistakes or misuse are not only possible, but arguably more likely. While we all pride ourselves on our care and attention to good practice, we have perhaps failed to apply sufficient scrutiny in the past to the downsides of our positive work. Today, we can no longer evade this.

Whether we like it or not, or agree with this viewpoint or not, the fact is that there are ever more opportunities 'out there' and within our own labs and offices that have the potential to cause or contribute to biosecurity problems. As guardians of science, we need to do something about this.

---

[10] *Ibid*, p. 16.

But My Work is Not Dangerous....

I am sure that most readers will, at this point, be patting themselves on the back and smiling. 'My work is not a threat to anyone.' I have never yet met a scientist who does *not* think this (even the former weapons scientists I have worked with didn't think their current work was a danger). But how do you know that? Have you looked at your work and actually *tried* to consider what harmful applications it could be put to? Do you *really* know that all your staff and students are above reproach and would never misuse your work, your teaching and the knowledge you are passing on to them, for adverse purposes?

This is just looking at *purposeful* misuse of your work. What about negligent or accidental adverse outcomes of your work? Are you sure that you have systems in place at work that will always prevent accidental adverse outcomes? Why do we have lab accident logs, if that is true? And are you sure that your biosecurity activities can also prevent all biosecurity problems? We all need to think about these things. I will give examples of some real-life cases later.

It is worth pointing out here that the commonly-held idea of a biological or chemical weapon is not the only one that is valid. Most of us would picture such a weapon as being some sort of container holding the biological or chemical agent attached to a missile, a bomb or a shell. However, this is only part of the reality. One of the aspects of biochemical weapons that makes them so insidious is the fact they do not need to be delivered by a conventional weapons system. A hostile person could effect the use of a biological or chemical weapon simply by adding an agent to a water supply, or spraying it in a public place, or tainting foods on a factory production line or supermarket shelf. You do not need a bomb or a missile to deliver a biochemical weapon. You can mount an attack using a perfume atomiser if you want to.

Taking Responsibility

Today, we cannot avoid accountability and responsibility. The public, our funders, our journal editors and others all demand it. Those who

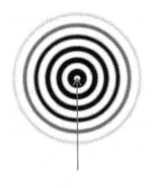

You and your actions
Your internal colleagues and their actions
Your external work contacts and their actions
Your family and friends and their actions
Colleagues' families/friends and their actions
Local population and their actions
Wider population (national)
Global population (international)

YOU ARE HERE

Figure 1.1: The Ripple Effect Diagram — where would you place the actions of each group of people after yourself, and where and when would the effects (the ripples) end?

do not yet, soon will. We must take increasing account of the nature and scope for misuse of our work given the advances we have made and the potential we now have for manipulating life itself.

Take a look at Figure 1.1 and consider your work and its effects. If you want your *good* work to 'spread out' and have an effect, you would welcome the ripple effect, in which one ripple causes the next. You are pleased when your work 'spreads'. But what if someone misuses your work, or carries out hostile work that your original work had pioneered or enabled — what would you think about the spreading ripples then? Unfortunately, it is not just you and your work that can cause problems — think of all the people with whom you come into contact while you are engaged in a project; they all have contact with you just as much as you have contact with them. Once you have thrown the pebble into the pond, where will the ripples stop spreading?

Most scientists are affronted when the 'biosecurity police' come along and say that their work can be misused for bad purposes. I know that I would react in the same way had I not been shown the nature and scope of the problem. However, the notion that scientific work can have adverse outcomes, intentional or not, has been around for a long time. There is nothing new about it. We are already familiar with the notion of side effects arising from the use of medicines. Some side effects are known at the point of commercialisation of a new drug, but are deemed worth the benefits. Physicians will prescribe these drugs but perhaps take some mitigating action to minimise the consequences of the side effects. For example, we know that aspirin is an effective analgesic, but that it can cause gastric problems. Some people may have to avoid aspirin altogether and use other forms of analgesia. But we don't abandon the use of aspirin because it adversely affects some people. We simply put in place certain caveats as experience and research highlight new side effects. Other drugs are deemed *safe* at the point of commercialisation, but prove to be problematic *later*. In the 1960s, the drug Thalidomide was widely prescribed for early-pregnancy 'morning sickness' and a range of other complaints. Only later did thousands of babies emerge with missing or malformed limbs. Granted, medical testing has moved on since the 1960s, but we can, in truth, never be utterly sure that a new drug is free from unintended consequences. We can say the same for many sorts of scientific research.

While these examples relate to some *unintended* adverse consequences of science, we need to be aware today, more than ever, that good science can also lead to knowledge that can be used for *intended* adverse outcomes of research. This is the paradox of dual use. As I said above, dual use awareness must therefore be part of every scientist's skill set.

Three Clear Examples of Dual Use — Potential and Actual

The Australian Mousepox Experiment[11]

This paradigm case of potential unintended dual use has been widely reported and discussed in academic journals, reports, books and in

---

[11] Jackson, R.J., Ramsay, A.J., Christensen, C.D., Beaton, S., Hall, D.F. and Ramshaw, I. (2001) Expression of Mouse Interleukin-4 by a recombinant ectromelia virus suppresses cytolytic lymphocyte responses and overcomes genetic resistance to mousepox. *Journal of Virology* 75(3), 1205–1210.

the science and mainstream media.[12] In 2001, researchers in Canberra, Australia, were carrying out work for a major social and economic benefit — the control of a serious mouse pest problem in Australia. Using standard genetic engineering techniques, they inserted the gene for interleukin-4 (IL-4) into the mousepox virus (ectromelia virus, ECTV). It was their hope that this would produce sterility, serving as an effective 'infectious contraceptive' in Australian mice.

However, a serious unexpected outcome occurred. The recombinant mousepox virus proved to be lethal to mice *which had been previously vaccinated* against mousepox. In other words, the manipulation of the mousepox virus by inserting IL-4, rendered the previously *non-lethal* mousepox virus *lethal*. This was not what was intended by the researchers. They immediately recognised that this was something that needed to be handled carefully, not only in literal terms in the laboratory, but in social responsibility terms.

Of particular concern was the possibility that the same techniques used to engineer the mousepox virus could be applied to create more virulent forms of poxviruses that afflict humans, including a vaccine-resistant strain of smallpox; one of the most devastating diseases in human history. Although it was eradicated in the 1980s, fears remain that former Soviet stockpiles — or genetically reconstituted forms of the virus — could be put to use by nefarious agents.[13]

Fine, some may say. Some mice died in a lab. Smallpox is already eradicated globally. Big deal.

---

[12] Nowak, R. (2001) Killer mousepox virus raises bioterror fears. *New Scientist*. Available on: https://www.newscientist.com/article/dn311-killer-mousepox-virus-raises-bioterror-fears/; BBC News (2001) *Mouse virus or bioweapon?* BBC World Service. Available on: http://www.bbc.co.uk/worldservice/sci_tech/highlights/010117_mousepox.shtml; Fidler, D. and Gostin, L. (2008) *Biosecurity in the Global Age: Biological Weapons, Public Health, and the Rule of Law*, Stanford, California, Stanford University Press, p. 42.

[13] Weir, L. and Selgelid, M.J. (2010) The Mousepox experience: An interview with Ronald Jackson and Ian Ramshaw on dual-use research. *EMBO Reports* 11(1), 18–24. Available on: http://www.ncbi.nlm.nih.gov/pmc/articles/PMC2816623/. Accessed 30/12/15.

However, it does not take much thinking to move this on:

> [Ian Ramshaw, one of the researchers, said] The critical time was the day that the vaccinated mice died. Ron Jackson [another of the researchers] came up from the animal house and said, "The vaccinated mice are dying." We just looked at each other and said, "Wow." We were now aware of something that hadn't been previously identified. I did not know of any other virus or system that could overcome a previously vaccinated regime.

This put a different slant on the research. Now it was seen to be possible to genetically manipulate a virus so that it overcomes the effects of a previously-given vaccine. Just how welcome might such a development be to someone who wanted to use such a virus as a biological weapon? Utilising the technology used in this experiment on mice in Australia, it would be possible to do something similar with human smallpox, for example. Or even, if desired, a pathogen could be manipulated to only affect certain genetic populations — an ethnicity-tailored weapon.

What did the researchers do next? Did they publish the full research and results?

> [Ian Ramshaw again...] We said, "Boy this is scary — this is the kind of thing that science fiction is made of." This was the first example of a virus overcoming vaccination, and this was very worrying. And I suppose there was a little bit of excitement about it as well — it wasn't all doom and gloom. This is exciting stuff, no matter how evil or bad it may turn out to be. We went away wondering what to do about it. In those times there was no pathway in the structure of scientific institutions for resolving a case like this. I gave a talk at a retreat when all our researchers were there. I gave them the results and asked them, "What do we do? Do we publish or don't we?" We came away with the consensus of the scientists, who probably weren't qualified, that there was already so much out there that could be used by bioterrorists that, I think I can quote, "One more won't make a difference". We informed the military and we never heard anything back. They probably wondered "Who the heck are these people?" or "What the heck is this?"

[Ron Jackson said] There is a history of efforts to eradicate smallpox and concerns about smallpox use in biological warfare. This was always in the background in our minds, although what we were doing wasn't in contravention of the Biological Weapons Convention [there is actually more to it than this, see below[14]] because it was for peaceful purposes and focused on animals rather than humans. But you would have to be an idiot not to realize the technology was transferable.

[Ian Ramshaw said] Now [2008] a long time after the original finding, I thought about this and realized there is another dual-use dilemma — and one that has not received so much attention. We created a transmissible virus that does not kill the individual but makes them sterile. That is as bad as making a virus that kills the individual. The principles were shown for mice; the principles were shown for rabbits; and there is no reason to think that similar principles would not apply to humans. I am only just realizing now [2008] that even before the so-called mousepox IL-4 experiments, *we were already undertaking 'dual-use' experiments.* [Italics mine].

Another issue was that you would never want to release a recombinant virus that you could not recover into the environment. No matter how many experiments you do to show that these viruses do not infect humans or other animals, there would not be sufficient clarity about the consequences of environmental release. So the original work should never have started in the first place. [I am not sure I would say that].

Is dual use starting to make more sense now?

---

[14] Jackson believed that what the team was doing was not in contravention of the BTWC 'because it was for peaceful purposes' — which it obviously was. However, the BTWC also states that member States (through their scientists) must not 'develop, produce, stockpile or otherwise acquire or retain' biological weapons (Article I); this obviously puts scientists in a quandary when they have inadvertently produced something that could be used as a biological weapon. Technically, such work should be stopped. But in reality, what actually happens? How can scientists meet their legal responsibilities to the BTWC while carrying out and retaining useful work? This is where discussions mediated by ethics can help in applying suitable mediating processes to contentious science in order to protect it and to protect the environment, people and animals at the same time. It needs to be said here that Ramshaw and Jackson acted in good faith, however, in 2001 when they did the work.

The Galston fertilizer case

It is worth looking as well at this example — an actual application of dual use. Who would think that a fertilizer could cause one of the great ethical and political debates of the 20th century? But it did. It is arguably right up there with the Manhattan Project in terms of the ethics of scientific progress.

Arthur W. Galston was a botany graduate in the US during the 1940s. He identified a synthetic substance that improved the growth of plants. As well as emphasising the positive aspects of his synthetic substance, Galston also reported, in his 1943 thesis, that if the substance were applied in heavy concentrations, it was toxic. Galston had developed a fertilizer that could also, when applied in excess, cause plants to shed their leaves. Later, without his permission, and unknown to him, his work was used in the development of Agent Orange, a powerful defoliant. This chemical substance was sprayed in massive amounts (around 20 million gallons) over Vietnam by the US government during the Vietnam War. The military aim was to reduce hiding places for the enemy and to help identify enemy movements and positions more easily. Unfortunately, Agent Orange (named after the orange bands around the containers in which it was kept) contained dioxins, which were later found to be associated with human birth defects and a range of other human health problems, thus having long lasting adverse effects beyond the period of intended use.

The effects of Agent Orange are still seen today, not only in those who were directly affected by exposure, but also in those of subsequent generations affected by birth defects and other impairments.[15] Galston said later:

> I used to think that one could avoid involvement in the anti-social consequences of science simply by not working on any project that might be turned to evil or destructive ends. I have learned that things

---

[15] King, J. (2012) Birth Defects Caused by Agent Orange. *Embryo Project Encyclopedia.* Available on: http://embryo.asu.edu/handle/10776/4202. Accessed 30/12/15.

are not that simple. . . . The only recourse is for a scientist to remain involved with it to the end.[16]

Galston died in 2008. He had been heavily involved in communicating the health risks of Agent Orange to the US government in the 1960s. His interventions resulted in the cessation of use of Agent Orange as a weapon in the Vietnam War. His career after the US military use of Agent Orange was marked by substantial work in bioethics. He had worked in good faith at his PhD research. It had useful social applications for agriculture that could bring great social and economic benefits to many. Unfortunately, it was also used as the basis for a biochemical weapon in a war that affected non-combatants as well as combatants and still has adverse effects today.

Is Galston responsible for the development of Agent Orange? If not, what degree of responsibility, if any, does he hold? If he is not responsible, who is? Should he have done something to prevent his work being misused in this way? Why should he have expected anyone to misuse his work? Was the development of Agent Orange a misuse *at the time*? Most reasonable people today would say that it was — but is this view simply benefitting from hindsight? Even had Galston wished to protest at the misuse of his work prior to it being actually used, on what grounds could he have complained at that time, given that the use was planned as a military tactic by his own government? What are the ethical responsibilities of the US government to those who were victims of Agent Orange both before and *after* concerns were raised about its effects on humans? Where does the responsibility stop?[17]

---

[16] Galston, A.W. (1972) Science and Social Responsibility: A Case History. *Annals of the New York Academy of Science* 196, 223. The quote given above is widely published online.

[17] Sture, J. (2014) Dual Use. In *Global Encyclopedia of Global Bioethics* (Springer). Available on: http://www.springerreference.com/index/chapterdbid/398716. Accessed 30/12/15.

Aum Shinrikyo — intentional dual use in Japan

In Japan, the religious extremist group Aum Shinrikyo was involved in several intentional attacks on the Japanese public. The first, which was largely ineffectual, involved the release of a liquid suspension of *Bacillus anthracis* which was aerosolised from the roof of an eight-storey building in Kameido, Tokyo in 1993. This resulted in reports to local environmental health authorities about foul odours and symptoms of gastro-intestinal disturbances among locals over the following days. The nature of this event was not recognised fully at the time, and it was not until the Sarin gas attack on the Tokyo subway in March, 1995 that the ability of Aum Shinrikyo to use harmful agents to cause mass fatalities was identified. Investigative work was carried out later, between 1999–2001, when tests were conducted on a liquid sample that had been collected on-site in Kameido at the time of the aero-solisation attack in 1993. From these tests, *B. Anthracis* was identi-fied. The samples obtained were found to be identical to a strain used in Japan to vaccinate animals against anthrax, which was consistent with testimony obtained from Aum Shinrikyo members about the strain source. No human cases were identified following the aerosoli-sation incident, and this was attributed to a number of factors includ-ing the use of an attenuated *B. anthracis* strain, low spore concentrations, ineffective dispersal, a clogged spray device, and inac-tivation of the spores by sunlight. Nevertheless, this incident raises certain questions. From what source did the group obtain the anthrax samples they used? Where did these activists learn their science? Could there have been any way of identifying them as ill-intentioned individuals with dangerous skills prior to the attacks?

The Tokyo sarin attacks took things much further. Aum Shinrikyo began producing sarin around 1993, testing it on sheep in a remote area of Australia (discovered later). They used this gas to mount sev-eral attacks, including an open-air night time 'test' attack in Matsumoto (1994) and in the subway attacks in Tokyo (1995). The Tokyo subway attacks are well documented elsewhere, but they resulted in multiple deaths and injury to many, with sarin being released in such enclosed public spaces. These incidents demonstrate the relative ease with which a few ill-intentioned individuals with the relevant expertise and equipment can carry out far-reaching attacks.

What Does this Mean for Scientific Practice?

Let us think about some questions that arise from the cases I have described here. Is it acceptable for research to continue once a potential or actual dangerous outcome is identified? I would say 'yes', depending on the context, but additional responsibilities need to be identified and extra safeguards need to be put in place. Let us think about some questions arising from this.

- At what stage of research should the search for dual use potential be started?
- At what stage of research should the search for dual use potential be abandoned? (That's a trick question).
- Who benefits from the research going ahead?
- Who decides what a benefit looks like?
- Who pays a cost for the research?
- Who decides what is a cost and what is a benefit?
- What counts as harm?
- To whom does it count as harm?
- Do they agree?
- Does it count as harm to anyone else?
- What will this harm actually look like?
- How will you know harm has occurred?
- What are the limits of responsibility and accountability of researchers who carry out scientific work?
- What *time limit* in responsibility would you apply?
- What *geographical limit* in responsibility would you apply?
- How do you know that you can identify the 'right' limits?
- If you are carrying out speculative, 'blue sky' research, what are the limits beyond which you would not go in terms of research?
- Do you think that scientists should be held accountable for misuse of their work, even if the misuse is carried out by someone else?

- How could you prevent misuse of your work by others?
- What would you do if you were sued, publically criticised or otherwise challenged over misuse of your work (even if not carried out by you)?
- To whom would you go for help?
- What would your defence be?
- If responsibility for biosecurity does not rest with you, with whom do you think it rests?
- If you win the Nobel Prize for the *good* outcomes of your work, is it right and fair therefore that you should be censured in some way for any *bad* outcomes of your work? (Just asking!).

How Can We Raise Dual Use Awareness and Make Appropriate Responses?

The biosecurity world has moved on since 2001 when Jackson and Ramshaw did their mousepox research. We have had 9/11, the US anthrax letters,[18] the inception of the National Science Advisory Board for Biosecurity (NSABB) in 2004[19] and the use of chemical weapons in Syria as a means of waging war.[20] Numerous heavy reports have been written, guidance formulated and

---

[18] Federal Bureau of Investigation. *Amerithrax or Anthrax Investigation.* Available on: https://www.fbi.gov/about-us/history/famous-cases/anthrax-amerithrax/amerithrax-investigation. Accessed 28/12/15.

[19] See Joliat, J.N. (2011) National Science Advisory Board for Biosecurity (NSABB) *Encyclopedia of Bioterrorism Defense.* Available on: http://onlinelibrary.wiley.com/DOI:10.1002/0471686786.ebd0177/abstract. Accessed 27/12/15. The NSABB is a US advisory committee that is managed by the US Office of Biotechnology Activities within the Office of Science Policy at the National Institutes of Health (NIH). Its purpose is to 'provide advice, guidance, and leadership regarding biosecurity oversight of dual-use research, defined as biological research with legitimate scientific purpose that may be misused to pose a biologic threat to public health and/or national security'.

[20] Organisation for the Prohibition of Chemical Weapons (OPCW) — United Nations Joint Mission (2014). *Thirteenth monthly report to the United Nations Security Council.* Available on: http://www.un.org/en/ga/search/view_doc.asp?symbol=S/2014/767. Accessed 28/12/15.

recommendations made (see Chapter 2). Some non-scientists, along with much of the public, think and say that certain experiments should never be done. Others, usually scientists, say that they have 'academic freedom' and 'scientific freedom' so they will continue go 'where the science takes them' (or wherever they can get funded). Some academics and scientists argue that there are sufficient safety features in place to counter any risks that could turn out to be a hazard to humans, animals or the environment.[21,22] It seems that the two 'sides' are far apart.

However, I would argue that the two sides do not need to be apart. The doubting side — the public, governments, funding bodies, interested parties, and so on — needs reassurance that scientists take 'outside' concerns seriously and will do something to respond to them that goes beyond shouting about scientific freedom. The science side needs reassurance that the lobbying of non-experts will not 'interfere' with 'their' work (who owns science is another issue).

By using an applied research ethics framework as a way to facilitate biosecurity, it is possible to provide for both of these required reassurances. Scientists can show to doubters and worriers *how* they are identifying and responding to the inherent risks of biotechnology and other forms of science by using an ethics framework (to provide effective biosafety and biosecurity/bio-chemical security). The worriers and doubters can *see* that their concerns are being listened to, their concerns are recognised and responded to and that scientists are not only listening but *acting* in response.

Both sides need to step back and respect the other side. Then we can all, hopefully, move forward together into a scientifically-improved world knowing that our concerns are being both recognised and addressed.

---

[21] Pinker, S. (2015) The moral imperative for bioethics *Boston Globe*. Available on: http://www.bostonglobe.com/opinion/2015/07/31/the-moral-imperative-for-bioethics/JmEkoyzlTAu9oQV76JrK9N/story.html. Accessed 30/12/15.

[22] Madhusoodanan, J. (2015) Bioethics accused of doing more harm than good. *Nature* 524, 139. Available on: http://www.nature.com/news/bioethics-accused-of-doing-more-harm-than-good-1.18128. Accessed 30/12/15.

Chapter 2

Dual Use Awareness as Part of Biological and Chemical Security

The Latest Fuss About Nothing?

If you were to write a book on some ancient, obscure subject that is likely to be read by only a handful of people, then perhaps you could argue that your work would have little or no adverse effect on society (although we all know the effects a book may have). If, however, you are working on some idea, material substance or process that will be 'used' in some way by, through, in, on or for humans, animals or plants, then what you are doing in the laboratory or at your desk *will* have an effect on society. After all, if your work is not going to have any effect outside the laboratory, why would you be doing it? Even theoretical 'blue sky' research is done with a view to future practical possibilities that may arise from it.

The question is, will that effect be, or could it be? Will it be beneficial, harmful, significant, inconsequential, frequent, rare, local or widespread? Will it affect one person or millions of people? Can it be reversed? Can it be recalled once it is 'out there'? Have you published all of your process and findings? Do you retain control of these after you have finished your part in the production of the work or publication? Who has access to the process and findings? Who could work out the process even if they are not involved in it with you? If you do not control it, who does? Does anyone control it at all? For how long?

Over what geographical area? What other uses are you aware of that your research could be put to that could be harmful? Do you have any idea of what your research may be used as a foundation for in the future? The list of potential questions is long.

I am not proposing here that you are responsible for all the outcomes, good or bad, that may ever be facilitated by your work and research. What I am saying is that in today's world of security challenges, you need to take thought about what adverse outcomes may be *reasonably foreseeable* at the time you are planning, carrying out and publicising your work. As well as taking this time and effort as an ongoing activity during your work (not just as a tick-box exercise at the beginning of the work), I also suggest that you should produce formal written evidence of how and when you have taken this consideration. If you are ever challenged on any adverse outcome, you can refer to this written evidence showing what consideration you gave to the risks, who you consulted, records of the discussions (such as minutes of meetings) and a clear decision-making process that underpinned what you ultimately did in terms of scientific work. For many, such a paper trail can be constructed around the usual ethics approval processes, but for others, especially those embarking on potentially dangerous or contentious work from the outset, it would be helpful to consult widely and to make a significant point of recording the ethical consideration of your work. You never know, those notes and minutes may keep you in your post one day or defend you in court. Your institution may also be protected against liability by these notes and records.

This brings me onto a further key point. I know that many scientists have to routinely get their work through ethics approval committees and so on. Many readers may actually be members of such committees. That is good, but in my experience, most such committees have little or no experience of dual use awareness or security knowledge. This is a good opportunity for you as a scientist to pass on your new or widened knowledge and awareness to ethics committees and approval boards. In this way, dual use awareness and biological and chemical security will be disseminated through official channels as well as unofficial, personal ones as individual scientists and small groups pick this knowledge up and incorporate it into their daily practice.

A comment I have often heard from science colleagues when I am working with them on promoting biological and chemical security is 'Why has all this suddenly started now?' Many scientists I have worked with have been tempted to think that in the past, when the pace of scientific development was often slower and less organised, scientists were forgiven for concentrating only on the work at hand with little or no thought for possible misuse of their ideas in the future. However, to see how long such concerns have been around, we need only to look at the earliest authorities to see how they themselves considered 'good' and 'bad' outcomes of scientific endeavour. The Hippocratic Oath must have developed to counter problems with medical interventions, otherwise what would have been the need for it? This oath is a good example of the early recognition of applied ethics, covering as it does a range of ethical behaviours expected of the medical practitioner. Amongst its original demands it made clear that the individual taking the oath would 'take care that they [patients] suffer no hurt or damage'.[1] This is not the place to go into the history of science ethics, but we can see that ethics in science is not a new development. Dual use awareness, enacted in and through biological and chemical security, is simply an extension of the ethics that already govern our practice as scientists.

Recent Policy and Professional Science Education Trends

Biosecurity responses post-9/11

In order to provide a simple summary of recent work and trends, it is useful to look at what has been happening in the USA and the UK since 2000, as the reports that emerged from these are (for good or ill) influential and relevant globally. I have purposely not pursued similar work from other countries, not because they are in any way

---

[1] The Hippocratic Oath in its original form was translated in a 19th century medical work available online. Copland, J. (ed) (1825). 'The Hippocratic Oath'. *The London Medical Repository Monthly Journal and Review* 23(135), 258. You can download this resource free of charge by searching for it on Google Books (the specific URL is too long to include here).

inferior, but because the purpose of this book is not carry out a global review of recent biosecurity work. For the sake of simplicity and brevity I will only refer to the UK and the US reports and to international treaties such as the Biological and Toxin Weapons Convention and the Chemical Weapons Convention. Interested readers can pursue further reading to complement this book if they wish.

Most current efforts appear to attempt to build on the traditions of self-governance in the life sciences to develop measures that will contribute to the 'web of prevention,' including:

- 'cradle-to-grave' approaches to oversight of research,
- Awareness-raising and education to inform and engage scientists to expand the existing culture of responsibility to include dual use concerns,
- independent input on the potential implications of trends in science and technology to inform governments and international organisations.

There is still a major focus on the perceived dangers of science and many reports take what McLeish and Nightingale[2] referred to as the *technology transfer* approach (Chapter 1). This means that many reports and the focus of much work emphasise efforts to stop 'them' getting hold of the 'bad' bits of science. As we've already agreed in Chapter 1, there are no 'bad' bits of science, only the potential for the hostile *misuse* of science.

Following 9/11 and the 2001 'Anthrax letters',[3,4] the USA began to seek a mix of measures that could mitigate the risks of any dual use

---

[2] McLeish, C. and Nightingale, P. (2007) Biosecurity, bioterrorism and the governance of science: The increasing convergence of science and security policy *Research Policy* 36, 1635–1654.

[3] See Federal Bureau of Investigation website (No date of publication available) *Amerithrax or Anthrax Investigation*. Available on: https://www.fbi.gov/about-us/history/famous-cases/anthrax-amerithrax/amerithrax-investigation. Accessed 27/12/15.

[4] National Research Council (2011) *Review of the Scientific Approaches Used During The FBI's Investigation of the 2001 Bacillus Anthracis Mailings*, Washington DC: National Academies Press. Available on: http://www.nap.edu/catalog/13098/review-of-the-scientific-approaches-used-during-the-fbis-investigation-of-the-2001-anthrax-letters. Accessed 27/12/15.

of benign scientific advances while enabling continued scientific progress. Crucially, it was recognized that scientific results should still be widely accessible to society. It was also recognized that there is a need to raise the awareness of dual-use risk among scientists, and to alert them to this new responsibility, at both national and international levels.

The Fink, Falkow, and Lemon–Relman Reports 2004–2006

In the following few years, a number of key reports were produced. The Fink Committee's *Biotechnology Research in an Age of Terrorism*;[5] the Falkow Committee's *Seeking Security: Pathogens, Open Access, and Genome Databases*[6] and the Lemon–Relman Committee's *Globalization, Biosecurity, and The Future of the Life Sciences*[7] all developed further ideas and recommendations on the current status of dual-use risks and how to counter them.

The Fink Report states that the "capacity for advanced biological research activities to cause disruption or harm, potentially on a catastrophic scale.....consists of two elements:

(1) the risk that dangerous agents that are the subject of research will be stolen or diverted for malevolent purposes; and (2) the risk that

---

[5] The 'Fink Report' — National Research Council Committee on Research Standards and Practices to Prevent the Destructive Application of Biotechnology (2004) *Biotechnology Research in an Age of Terrorism* Washington DC: National Academies Press. Available on: http://www.nap.edu/catalog/10827/biotechnology-research-in-an-age-of-terrorism. Accessed 27/12/15.

[6] The 'Falkow Report' — National Research Council Committee on Genomics Databases for Bioterrorism Threat Agents (2004) *Seeking Security: Pathogens, Open Access, and Genome Databases.* Washington DC: National Academies Press. Available on: http://www.nap.edu/catalog/11087/seeking-security-pathogens-open-access-and-genome-databases. Accessed 27/12/15.

[7] The 'Lemon–Relman Report' — National Research Council Committee On Advances In Technology And The Prevention Of Their Application To Next Generation Biowarfare Threats (2006) *Globalization, Biosecurity, and the Future of the Life Sciences.* Washington DC: National Academies Press, p. 28. Available on: http://www.nap.edu/catalog/11567/globalization-biosecurity-and-the-future-of-the-life-sciences.

Table 2.1:   Summary of the Recommendations of the Fink Committee 2004 (pp. 112–126 of the Fink Report)

Recommendation	Explanation
R1 Educating the Scientific Community	We recommend that national and international professional societies and related organizations and institutions create programs to educate scientists about the nature of the dual use dilemma in biotechnology and their responsibilities to mitigate its risks.
R2 Review of Plans for Experiments	We recommend that the Department of Health and Human Services (DHHS) augment the already established system for review of experiments involving recombinant DNA conducted by the National Institutes of Health to create a review system for seven classes of experiments (**the Experiments of Concern — see Table 2.2)**) involving microbial agents that raise concerns about their potential for misuse.
R3 Review at the Publication Stage	We recommend relying on self-governance by scientists and scientific journals to review publications for their potential national security risks.
R4 Creation of a National Science Advisory Board for Biodefense	We recommend that the Department of Health and Human Services create a National Science Advisory Board for Biodefense (NSABB) to provide advice, guidance, and leadership for the system of review and oversight we are proposing.
R5 Additional Elements for Protection Against Misuse	We recommend that the federal government rely on the implementation of current legislation and regulation, with periodic review by the NSABB, to provide protection of biological materials and supervision of personnel working with these materials.
R6 A Role for the Life Sciences in Efforts to Prevent Bioterrorism and Biowarfare.	We recommend that the national security and law enforcement communities develop new channels of sustained communication with the life sciences community about how to mitigate the risks of bioterrorism.
R7 Harmonized International Oversight.	We recommend that the international policy-making and scientific communities create an International Forum on Biosecurity to develop and promote harmonized national, regional, and international measures that will provide a counterpart to the system we recommend for the United States.

the research results, knowledge, or techniques could facilitate the creation of 'novel' pathogens with unique properties or create entirely new classes of threat agents."[8]

The Fink Report contained seven recommendations to ensure responsible oversight for biotechnology research with potential

Table 2.2: Definitions of 'experiments of concern' according to the Fink Committee 2004 (pp. 114–115 of the Fink Report)

Type of Experiment	Examples
1 Would demonstrate how to render a vaccine ineffective.	This would apply to both human and animal vaccines. Creation of vaccine resistant smallpox virus would fall into this class of experiments.
2 Would confer resistance to therapeutically useful antibiotics or antiviral agents.	This would apply to therapeutic agents that are used to control disease agents in humans, animals or crops. Introduction of ciprofloxacin resistance in *Bacillus anthracis* would fall into this class.
3 Would enhance the virulence of a pathogen or render a non-pathogen virulent.	This would apply to plant, animal, and human pathogens. Introduction of cereolysin toxin gene into *Bacillus anthracis* would fall into this class.
4 Would increase transmissibility of a pathogen.	This would include enhancing transmission within or between species. Altering vector competence to enhance disease transmission would also fall into this class.
5 Would alter the host range of a pathogen.	This would include making non-zoonotic into zoonotic agents. Altering the tropism of viruses would fit into this class.
6 Would enable the evasion of diagnostic/detection modalities.	This could include microencapsulation to avoid antibody-based detection and/or the alteration of gene sequences to avoid detection by established molecular methods.
7 Would enable the weaponisation of a biological agent or toxin.	This would include the environmental stabilization of pathogens. Synthesis of smallpox virus would fall into this class of experiments.

[8] Fink Report p. 1.

bioterrorism applications. One of these recommendations was to create a National Science Advisory Board for Biosecurity (NSABB) within the Department of Health and Human Services to provide advice, guidance, and leadership for a system of review and oversight of experiments of concern.[9] The National Science Advisory Board for Biosecurity was duly created and chartered in March 2004. The responsibilities of NSABB include many of the recommendations suggested by the National Academies; its charter outlines 12 responsibilities covering the identification of research of concern, the education of life scientists, regulation of scientific work, codes of conduct, and policy and advisory activities.

The Falkow committee specifically sought and considered the views of working scientists as it assessed the dual-use risks of the burgeoning genome-related research field after the mapping of human and other genomes in recent years. Given the existence of genome databases and the existing range of genomic data freely available at that time on the internet and in other sources, the Committee on Genomics Databases for Bioterrorism Threat Agents and the National Academies held a workshop in 2003 focussing on the identification of threats and the implications of and for the public availability of genomic data.

The ensuing Falkow Report (2004) included four recommendations. These were that firstly, the public availability of genomic data on microbial pathogens should not be restricted but should continue and be encouraged, as current governance policies were effective and restriction of data was not practical. Secondly, that genomics and genome sequence data should be exploited fully to improve our ability to defend against infectious agents of all types, including those that contribute to human disease, as well as those naturally occurring or genetically enhanced organisms that could be used in

---

[9]For useful information, see: Shea, D. (2007) *Oversight of Dual-Use Biological Research: The National Science Advisory Board for Biosecurity* Congressional Research Service Report, Washington DC. Available on: https://www.fas.org/sgp/crs/natsec/RL33342.pdf. Accessed 27/12/15.

a bioterrorist attack. Thirdly, that advances in genome science should be regularly reviewed to keep all relevant government departments and agencies appraised of new developments that may affect national security. To implement this, it was recommended that regular meetings between the scientific and security communities should be held to discuss the implications of new developments and to develop 'coherent' responses. The use of the word 'coherent' here exemplifies the different discourses (ways of communication including terminology, focus, risk-perception, and so on) utilised by the scientific and security communities. Fourthly, the report endorsed Recommendation 7 of the Fink Report (2004), which called for an international forum to unify the discussion on the effect of genomics on biosecurity.

Finally, the Falkow report also endorsed the Fink Report's Recommendation 1, that called for national and international professional societies and related organizations to work to educate scientists about the risks of life-science research results being misused and about scientists' responsibility to mitigate these risks. The Falkow report commented that:

'Life-science research is global, and no single nation can successfully implement policy concerning access to and release of life-science data and results without reference to the rest of the international community. For that reason, it is of the utmost importance that the international community establish a common understanding of security concerns and shared resources to make the most efficient and safest use of genome data and experimental results, some of which might suggest how pathogens could be successfully enhanced. ..... If conducted openly and in the proper spirit, the process of discussing these issues might actually build understanding, and some trust, among the nations involved and eventually help to establish an international norm against misuse of genetic information'.[10]

---

[10] Falkow Report pp. 62–63.

Since the publication of the Falkow Report, an International Forum on Biosecurity has been established and has held a number of meetings and produced publications.[11]

The Lemon–Relman Report (2006) focussed on those advances in the life sciences, related and convergent technologies that were deemed likely to alter the biological threat spectrum over the next 5–10 years. The report took a more global approach than had the Fink Report. The Lemon–Relman Committee's charge was to examine current trends and the future objectives of research in the life sciences that may enable the development of a new generation of future biological threats. They made recommendations based on their assessment of these trends and objectives. Their conclusions reflect a fundamental judgment that 'the future is now' and that time is pressing.[12] This led to the conclusion that, given continuing and rapid advances, the task of surveying current technology trends in order to anticipate what new threats may face the world in the future will be never-ending. The report suggested that it is increasingly important that life scientists take steps to ensure that their work should not be exploited through dual use, and that this requires that those working in the life sciences achieve a much greater appreciation of the dangers than was held at the time of the report by most, plus a greater willingness to shoulder this responsibility. As the committee's co-chairs argued in their preface, 'A new ethos is required, and it must be achieved on a global scale...'[13]

The report endorsed and affirmed policies that, to the maximum extent possible, should promote the free and open exchange of information in the life sciences. It also recommended the adoption of a

---

[11] Committee on International Outreach Activities on Biosecurity, Board on International Scientific Organizations, Policy and Global Affairs and National Research Council of the National Academies (2009) The 2nd International Forum on Biosecurity: Summary of an International Meeting, Budapest Hungary, March 30–April 2, 2008. Washington DC: National Academies Press. Available on: http://www.nap.edu/search/?term=The+2nd+International+Forum+on+Biosecurity%3A%3A+Summary+of+an+International. Accessed 28/12/15.

[12] Lemon–Relman report p. viii.

[13] Lemon–Relman report p. ix.

broader perspective on the 'threat spectrum' and the strengthening and enhancing of scientific and technical expertise within and across security communities. The report also recommended the adoption and promotion of a common culture of awareness and a shared sense of responsibility within the global community of life scientists and a strengthening of public health and existing response and recovery capabilities. Further, it suggested that scientists and others should adopt a broadened awareness of threats beyond the classical 'select agents' (officially recognised and controlled agents) and other pathogenic organisms and toxins, so as to include, for example, approaches for disrupting host homeostatic and defence systems and for creating synthetic organisms.[14]

More recently, specific challenges have been addressed in meetings and reports. Following the international controversy surrounding the work on H5N1 carried out in the Netherlands and the US in 2011–2012, a public workshop was held to discuss the issues raised by this and similar research and to provide a forum for airing the concerns of a wide range of bodies and areas of society. The final report on this workshop[15] included some useful questions and considerations around the dual use debate. It also included a report on a useful session held at the workshop that focused on 'Scientists and the Social Contract',[16] which covered public participation in science regulation and the concerns of scientists and the wider public. The report includes a range of questions posed at the workshop around regulatory mechanisms, their possible nature and scope, the extent of these across science and technology, what decision-making processes could be implemented, who would be involved and so on. It also considered the role of universities as educators and promoters of ethics in science and raised the issue of

---

[14] Lemon–Relman report p. 235.
[15] National Research Council (2013) *Perspectives on Research with H5N1 Avian Influenza: Scientific Inquiry, Communication, Controversy. Summary of a Workshop.* Washington DC: National Academies Press. Available on: http://www.nap.edu/catalog/18255/perspectives-on-research-with-h5n1-avian-influenza-scientific-inquiry-communication. Accessed 27/12/15.
[16] *Ibid*, p. 29.

public involvement in the decision-making processes involved. This workshop report is not only a useful guide to the H5N1 debate, but also a good example of the kinds of questions that must be asked in order to achieve appropriate levels of security in science. These include issues of awareness-raising, better ethics teaching and knowledge and mechanisms to increase public confidence in scientific activities.

International science meetings and workshops 2004 onwards

These reports, and others, have clearly highlighted the need for national and international scientific organizations to recognize education as fundamental to the effort to protect science from misuse. A primary role for these organizations is in providing endorsements for education that are respected within the scientific community, so that concerns about potential security risks should not be addressed simply in the form of policies, restraints, and obligations handed down from governments. The reports and other activities already mentioned also make clear that this is increasingly a global issue, and not one restricted to the US and the UK.

The educational theme has been endorsed by a number of key international science organizations. In 2004, the IAP: The Global Network of Science Academies[17] set up its Biosecurity Working Group (BWG)[18] to provide a forum for scientists and associated groups to discuss and agree the development of responses to security risks, particularly in the life sciences. In December 2005, the IAP made a Statement at the Meeting of the States Parties to the Biological and Toxin Weapons Convention,[19] presenting guiding principles that should be addressed in formulating codes of conduct for scientists.

---

[17] The Inter-Academy Panel — IAP: Global Network of Science Academies is, as the title indicates, a network of worldwide science academies formed in 1993 to advise on global issues from a scientific perspective. Available on: http://www.interacademies. net/About/18190.aspx. Accessed 27/12/15.

[18] For the IAP webpage on *Biosecurity*. Available on: http://www.interacademies. net/Activities/Projects/17806.aspx. Accessed 27/12/15.

[19] See the IAP Statement on Biosecurity. Available on: http://www.interacademies. net/10878/13912.aspx. Accessed 27/12/15.

These principles covered awareness of the problem of risk; safety and security; education and information; the accountability of scientists and the oversight of the risks and management of them of and by scientists and others. The IAP Statement on Biosecurity was endorsed by over 70 (out of 107) member academies of the IAP in 2005. Since then, IAP member academies have hosted a major meeting in Beijing in 2010 looking at 'Trends in Science and Technology Relevant to the Biological and Toxin Weapons Convention'.[20]

The International Council for Science (ICSU) has also made statements about the need for enhanced education and dual use awareness-raising. The ICSU Executive Committee endorsed education and awareness raising initiatives on dual use issues by its unions in April 2005 and also contributed to the first International Forum on Biosecurity held in Como, Italy, in March 2005 in collaboration with the IAP. A further ICSU meeting, focusing on science developments relevant to the BTWC, was hosted in London by the Royal Society in 2006. The report[21] arising from this event focused in part on education in science and endorsed the use of codes of conduct, although it also recognised the limitations of these.

The involvement of science unions and similar associations is key to enhancing awareness and developing educational strategies and policies to counter the risk of dual-use. A number of unions cover the 'life sciences' and it is clear that the chemistry sector and its unions are also key players in the debate, given the increasing areas of chemistry/biology convergence in research and industry. Two science unions, the International Union of Biochemistry and Molecular Biology[22] and the International Union of Microbiological Societies[23]

---

[20] For the report of this meeting. Available on: http://www.interacademies.net/Activities/Projects/17806/17890.aspx. Accessed 27/12/15.

[21] Report of the Royal Society, London, meeting. Available on: http://www.icsu.org/what-we-do/projects-activities/archived-projects-and-activities/biosecurity/pdf/BTWC_Workshop_Report.pdf. Accessed 27/12/15.

[22] International Union of Biochemistry and Molecular Biology code of ethics. Available on: http://iubmb.org/about-iubmb/code-of-ethics/. Accessed 27/12/15.

[23] International Union of Microbiological Societies code of ethics. Available on: http://www.iums.org/index.php/code-of-ethics. Accessed 27/12/15.

created codes of conduct/ethics after the 2005 BTWC intersessional process. Many now have activities related to aspects of science education into which dual use issues could be inserted. Moreover, the International Union of Pure and Applied Chemistry (IUPAC) in collaboration with Organisation for the Prohibition of Chemical Weapons (OPCW) carried out work in 2005–2006 looking at ways in which security concerns could be addressed through the education of chemists and application of ethics-based codes of conduct.[24]

An example of inter-union and inter-academy activities was the Second International Forum on Biosecurity, which was held in March 2008, hosted by the Hungarian Academy of Sciences.[25] The event was co-sponsored by the IAP along with InterAcademy Medical Panel (IAP) plus three major unions: the International Union of Biochemistry and Molecular Biology (IUBMB), the International Union of Biological Sciences (IUBS) and the International Union of Microbiological Societies (IUMS). This meeting was supported by several significant sources including the Alfred P. Sloan Foundation, the Carnegie Corporation of New York, the IAP, and the IUBMB, indicating the widespread concern of organisations and funding bodies about the topic. The meeting focused on education, research oversight, and roles for international scientific organizations.

A further workshop was held in November 2009 on *Promoting Dual Use Education in the Life Sciences*. This event was also sponsored by the IAP, IUMS, and IUBMB, and was hosted in Warsaw by the Polish Academy of Sciences, supported by the U.S. Department of State, and the IAP. Over 60 participants from 25 countries, along with colleagues from UNESCO, and the Implementation Support Unit of the BTWC attended the meeting. Participants included

---

[24] For International Union of Pure and Applied Chemistry (IUPAC) responses. Available on: http://media.iupac.org/publications/ci/2005/2703/pp3_2004-048-1-020.html and http://www.iupac.org/nc/home/projects/project-db/project-details.html?tx_wfqbe_pi1%5bproject_nr%5d=2004-048-1-020. Accessed 28/12/15.
[25] Report of the Budapest meeting in 2008. Available on: http://www.nap.edu/catalog/12525/the-2nd-international-forum-on-biosecurity-summary-of-an-international. Accessed 28/12/15.

experts in biosecurity, experts in bioethics and the responsible conduct of research from academia, NGOs, and governments. Added to these experts were representatives from a new element — experts in the 'science of learning,' that is, experts on what research tells us about how people learn at different stages of life and what this means for effective approaches to teaching. The international steering committee under the auspices of the US National Academy prepared the report of the Warsaw meeting, which was released in mid-September 2010.[26]

Responses from Funding Bodies and Journal Editors

In 2015, a Working Group that had been convened in 2013 to revise a joint statement by British research funding bodies, released a revision of their original 2005 statement. The Wellcome Trust, the Medical Research Council and the Biotechnology and Biological Sciences Research Council (BBSRC) clarified and updated their 2005 statement to include five steps for scientists to take in applying for funding from these bodies to carry out research. These are:

- inclusion of a question on application forms requiring applicants to consider short and medium-term risks of misuse associated with their proposal,
- explicit mention of risks of misuse in guidance to external experts who peer review grant applications as an issue to consider,
- development of clear guidance for funding committees on this issue and the process for assessing cases where concerns have been raised,
- modification of guidelines on good practice in research to include specific reference to risks of misuse,

---

[26] Committee on Education on Dual Use Issues in the Life Sciences (2010) *Challenges and Opportunities for Education About Dual Use Issues in the Life Sciences*. Washington DC: National Academies Press. Available on: http://www.nap.edu/catalog/12958/challenges-and-opportunities-for-education-about-dual-use-issues-in-the-life-sciences. Accessed 27/12/15.

- a requirement to notify funders, and other relevant authorities, of any change in the status of a dual use risk or any new risks that may emerge during the course of a funded research project that were not anticipated in the original application.[27]

This is clearly a tight, highly-focused approach to the oversight of work funded by these three major UK funding councils. Of course, in order for such requirements to be met effectively by researchers, suitable and appropriate training in the recognition of dual use risks is a necessity. Have you had such training? Do you agree that such training needs to be a vital part of the skill set of all scientists?

The journal *Nature* now has a policy on biosecurity and materials submitted to it for publication,[28] with advice for authors and clear information about how the editors will seek technical advice if they deem it necessary when submitted papers and other materials require this. *Science,* on the other hand, expects authors to bring to the attention of its editors any submitted work that may be of dual use concern,[29] in which case it will pass this on to the Editor in Chief for further consideration. These two policies are therefore not quite the same, with the *Nature* policy implying that its editors actively seek out security concerns as part of the review process, whereas the *Science* advice seems to indicate that the problem is left to authors to highlight in their submissions — presumably avoiding the need for editors to review submissions for security risks themselves. Either way, these two examples highlight the fact that the two most well-known global science journals have recognised the issue of security in science and are seeking to pass this awareness on to authors. This is a big step in itself. How do you meet these requirements when you submit work to these two journals?

---

[27] Wellcome Trust website: Available on: http://www.wellcome.ac.uk/About-us/Policy/Policy-and-position-statements/wtx026594.htm. Accessed 28/12/15.

[28] *Nature* policy on biosecurity. Available on: http://www.nature.com/authors/policies/biosecurity.html. Accessed 28/12/15.

[29] *Science* security advice for authors. Available on: http://www.sciencemag.org/site/feature/contribinfo/prep/gen_info.xhtml#security. Accessed 28/12/15.

Table 2.3: Findings and conclusions of the 2010 Bradford meeting. *Dual use Education for Life Scientists: Mapping the Current Global Landscape and Developments* [30]

Global issues in education around biosecurity

A lack of access to effective teaching materials, or to materials relevant to diverse audiences, and to materials in languages other than English.

A lack of faculty/trainers prepared to teach about dual use issues or who are aware of strategies for such teaching, and a lack of opportunities to receive such training.

A lack of standards for designing and assessing courses/modules/materials and of networks to share best practices and lessons learned.

A lack of resources and priority-identification for the inclusion of dual use education as part of training for life scientists.

A lack of "champions" to help raise this priority and to support the provision of such resources.

Preparation

It is clear from the numerous surveys carried out in diverse global settings that there is a widespread lack of knowledge of dual use and biosecurity issues among the education and scientific communities; moreover, there are elements of hostility towards the issues from some quarters.

It is also clear from these surveys that once the topic is introduced, interest is aroused and participants are keen to develop their knowledge and skills in this area; this is particularly supported by the presence and activity of motivated, enthusiastic individuals.

It is necessary to support stakeholders in moving away from the ethics "tick-box" or "once-and-done" approaches to dual use bioethics and to encourage the notion of ethics as an ongoing activity and theme pervading all scientific practice.

The survey-follow up model appears to be successful in assessing need and engaging interest in specific countries, which allows for nation-specific development of dual use education packages.

It is clear that the delivery of dual use bioethics and biosecurity courses may be incorporated into existing curricula focusing on biosafety and the responsible conduct of research.

(Continued)

[30] Sture, J.F. and Minehata, M. (2010) Dual-use education for life scientists: mapping the current global landscape and developments. Japan Society for the Promotion of Science -Economic and Social Research Council Seminar Series. Research Report for the Wellcome Trust Project on 'Building a Sustainable Capacity in Dual-use Bioethics'. Available on: http://www.worldscientific.com/worldscibooks/10.1142/q0027#t=suppl

Table 2.3:   (*Continued*)

Those involved in biosecurity and bioethics education for life scientists must be clear that life scientists already worked in an ethical manner and are already ethically-engaged; the introduction of dual use issues is simply another approach through which to consider their work.

There must be clear recognition that science does good and should not be demonised.

There is as yet no identified theory of dual use or of dual use bioethics or dual use biosecurity; this is needed in order to act as a sound foundation when building global models of education and action in the security and industrial sectors.

Such a theory and conceptual model need to take account of the diversity of cultural patterns of engagement with dual use, bioethics, and biosecurity.

Implementation

No "one size fits all" approach will be successful when implementing educational policy and practice.

The specific needs of each country or region must be accommodated in any course or set of resources, including, wherever possible, use of local languages.

It is important to involve local experts in the development of courses and resources to encourage a sense of ownership and to ensure the validity and reliability of educational activities.

Online education as well as face-to-face teaching and learning can work effectively, both in the student and professional development contexts.

The work pioneered by the Bradford Disarmament Research Centre in the form of the freely-available online Educational Resource Module (EMR) goes a long way towards meeting the needs of students, tutors and practising scientists for easily-accessible materials and opportunities to tackle the issue of dual use in educational and professional development settings. [The website for this material is no longer available online. You may be able to access the EMR by contacting the Division of Peace Studies at the University of Bradford.]

The EMR is already translated into several languages with more being planned; BDRC has evidence that educators around the world are using the resource and adapting it into their own courses and languages in whole or in part.

The Bradford Train-the-Trainer accredited online module, with its capacity to teach students via the web using various software packages has emerged as a powerful tool in sustainable capacity building and professional development for any scientist, educator, policy-maker or associated professional with a role to play in the biosecurity and bioethics stakeholder communities.

(*Continued*)

Table 2.3: (*Continued*)

Codes of Ethics and Codes of Conduct may be useful tools for the promotion of dual use awareness and the management of science with potential dual use application by scientists.

It is important that the professional science community and its stakeholders engage effectively with the amateur biologist community, in order to promote the concepts of dual use and biosecurity beyond the boundaries of higher education and industry.

Monitoring and Reporting

There is a need for clarity in defining roles and responsibilities in this area.

It is vital that scientists take a leading role in this process before others take that role for them.

It is vital that effective assessment/evaluation processes are developed with which to track progress in terms of uptake of dual use education and resulting changes in practical activities.

There are potentially many stakeholders, civil and military, private and public, involved in dual use-related responsibility, but it is clear that there is little agreement internally in any government as to where the lines of responsibility fall.

It is therefore necessary to clarify a model of ownership of dual use bioethics and biosecurity to ensure that these are engaged with effectively.

The role of professional associations is key in promoting, regulating, advising, monitoring, and reporting of educational packages and advances in the dual use context.

The role of journal editors and funding bodies is also key, as these largely dictate the opportunities for research to take place and to be disseminated.

It is necessary to build effective links between educational institutions, professional associations and national governments in order to set up appropriate monitoring and reporting mechanisms.

These must ultimately involve reporting to national and international bodies including the BTWC.

Expert-level biosecurity education meetings in Bradford, UK, 2010 onwards

There is clearly a need for opportunities for scientists, policy makers, and the security community to meet to discuss these issues. Since 2010 a number of meetings have been held at the University of Bradford, UK, providing a forum for biosecurity experts and 'new users' of biosecurity approaches to attend and share their knowledge and best practice. The first of these was funded jointly by the UK's Economic and Social Research Council and the Japan Society for the Promotion of Science, bringing together experts from the UK, the US, Japan, Sweden, Ukraine, Italy, and the Netherlands. During the two-day meeting work from around the world on biosecurity education (and the lack of it) was presented and discussed.[31]

A number of interesting and informative findings were compiled at the end of the July 2010 Bradford meeting. These are listed in Table 2.3 below. Further meetings were held in 2012 and 2013 to build on the momentum achieved by the 2010 meeting. The 2012 meeting focused on the lessons that could be shared about applied ethics between engineering and the life sciences[32] and the 2013 meeting focused on the ethical and security issues raised by the increasing convergence of chemistry and biology.[33] Crucially at the 2013 meeting, colleagues from developing countries who had experienced new biosecurity education initiatives funded by the UK were able to attend and share their views and experiences. This sort of sharing opportunity has to be a key priority in the sharing of best practice and advancement of biosecurity awareness and competency. A key finding that has emerged from each of these meetings is the need to widen science education to incorporate biosecurity issues into existing biosafety and bioethics provision and to make this a regular, mandatory element of continuing professional development (CPD) courses, in order to link it to professional competency standards. Indeed, a biosecurity competency standard is a key need for all the scientists if we are to counter the security risks that face us in the early 21st century. It can be seen from these findings that action is needed

---

[31] *Dual use Education for Life Scientists: Mapping the Current Global Landscape and Developments.* An international meeting hosted by the University of Bradford, UK, July 2010.

[32] *Yearbook of Biosecurity Education* (2012).

[33] *Yearbook of Biosecurity Education* (2013) forthcoming.

in the science community. For international readers, the *National Series*, is available from the author. This is a set of country-specific educational packages that are ready-to-be-delivered in a range of countries.[34]

The Dual Use Problem: How Does It Affect Me?

The Ethics Toolkit simply presents a new perspective on good research practice as a means to enhance your skills of recognition and effective response to biological and chemical security issues. On completion of this book, you will be on your way to being equipped with a set of skills and knowledge that complements your existing expertise in biosafety. You can employ these skills as a matter of course in your everyday scientific practice. By applying the ethical principles of autonomy, doing no harm, doing good, and so on, to scientific work, you will be able to identify dual use risks more easily. The application of the Toolkit will enable you to broaden your perspective beyond the laboratory door. Hopefully, you will feel able to pass this perspective and these skills on to your students, research assistants and colleagues. None of this ethical practice is difficult and it need take no longer to implement than biosafety measures already do. Once you are used to the principles and the questions within the Toolkit format, you will no longer have to go through each and every question, as you will be able to quickly spot areas of concern in any science project and employ the questions necessary for that.

I refer to a concept in the Toolkit that I have called 'the dual use context.' By this, I mean any situation in which you as a science

---

[34] Sture, J.F. and Minehata, M. (2011–2013) The *National Series*. This is a set of biosecurity-based lectures and notes for teaching or personal learning. Editions of the Series are provided for: Pakistan, Ukraine, Georgia, Tajikistan, Azerbaijan, Egypt, Jordan, Saudi Arabia, Iraq, ... Sture was funded by the UK and US governments to carry out this work. Available on: http://www.worldscientific.com/worldscibooks/10.1142/q0027#t=suppl

professional, or any of your students or associates, are involved in research or other work where you or they:

- Are assessing your own or another's work for possible dual use potential, or
- Have already recognised some dual use application(s) in your own or another's work and are planning and taking appropriate risk minimising activities.

While this book focuses on biological and chemical security in science, the principles can, of course, be applied to any scientific, engineering or other research and development context.

You will have a chance later to come up with some of the possible types of risk-minimising activities that you and your associates may take in order to enhance biological and chemical security, within an applied ethics framework. You will be able to do this by thinking about the questions posed at the end of each section in Chapters 5 and 6. There are few clear-cut answers and I do not pretend to offer 'black and white' solutions here — you know your own work best, and you are best-placed to assess and respond to the possible or actual adverse outcomes of it. In terms of looking at the work of others, I am not advocating that you 'spy' on other peoples' work, but that, through the adoption of an ethics approach using The Ethics Toolkit, you will gain a dual use perspective that will enable you to recognise potential or actual dual use problems as a matter of course in any work that you look at (just as you can already spot potential or actual biosafety problems easily).

Is This Social Science Rather than 'Real' Science?

Many of the ethical principles or approaches covered here are traditionally associated with the social sciences rather than the natural sciences. However, in adopting this wider ethical perspective, you are not going to be asked to do anything that you would be unfamiliar with in everyday life. Being a scientist gives you a certain analytical outlook on the world, but this does not diminish the ethical approach to interaction with other people that you already practice, even unwittingly, on a daily basis.

In practice, this approach to managing ethical issues comes down to the interaction or direct contact you have *with people*. In the laboratory setting this is largely the lab staff, researchers, technicians, administrators, visitors, and so on. While the wider public is of course already considered by scientists, it has not been necessarily a focus of your ethical deliberations to date. This is where dual use awareness comes in, as this approach requires that public health, safety, and security should take as prominent a position in your work planning as do biosafety and other forms of bioethics. This biological and chemical security approach simply addresses the potential for security issues to arise out of biotechnology and chemistry research; it applies to a wider context than that of biosafety, which although protecting the public indirectly, naturally tends to focus more on local containment of biochemical agents. Biological and chemical security, by contrast, is also concerned with the 'containment' or protection of ideas and information as well as biological and chemical agents. It looks 'beyond the laboratory door' as a matter of course, from the beginning of all work.

How Will the Ethics Toolkit Help?

In order to achieve effective dual use biosecurity, it is not intended that The Ethics Toolkit should tell you what to do in all research circumstances. It is intended, however, to outline the key ethical issues around biosecurity thus enabling you to apply them in a potential dual use context. This will help you to maximise your accountability for good ethical practice not only in relation to your own research but also that of your students or, in some cases (as PI) of your colleagues, where appropriate. It will certainly help you to recognise potential dual use opportunities in the work of others. By using the Toolkit, you will be engaging in a form of 'social responsibility ethics'. This facilitates effective biosecurity standards and enhances your biosecurity competence. It will also help to protect your employers from accusations of allowing 'dangerous' science to go ahead.

I am not concerned here with the philosophical background to ethics, but with a focus on the practical identification of potential dual use problems in the scientific research context. I am not going to tell you what the dual use risks are or may be within *your* science work (that is a task for you) — but the Toolkit will highlight how dual use risks may have an impact on you, your team, people associated with your work indirectly and the wider public. This is essentially about relationships between people and your work, plus the effects your work may have if it is misused, either intentionally or unintentionally.

When working in a laboratory setting it is easy to lose sight of the people who are 'end-users' of the results of research work, those who may come into contact with it, those who will be marginally affected by it, or those who may be affected by *someone else* using your research methods, materials or findings to carry out work for hostile purposes. The hostile nature of the dual use of biotechnology has the potential to affect millions of people just as it may affect only one person. As a scientist, you are already familiar with activities and expectations relating to the safety of yourself and your colleagues in and around laboratories. To supplement this, The Ethics Toolkit provides you with another, wider lens through which you can view your own work and that of others, looking for potential dual use flashpoints that need to be addressed.

Ethical issues and dual use risks may be found at the planning stage, the grant application stage, the recruitment stages, other materials- and information-gathering stages, the data collection and analysis phases, revision phases, the publication/dissemination phase, the testing phases or the 'end-user' stage, where products of your research, or outcomes of it, are launched into society. Given what we have seen already in the cases of Galston and his fertilizer and Ramshaw and Jackson's experiences with their mousepox research, we also need to add the 'unexpected outcomes' stage as well.

It as well to be aware that ethics affects all of our daily tasks, not simply when we are working at the bench or publicising our work. You will find that even when you are 'only' working with computers, laboratory equipment or doing the paperwork, you still need to identify and address any possible ethical, biological and chemical security issues. The following chapters will help you to consider a wide-lens perspective beyond your own immediate work, considering such issues as the effect your work may have on certain groups of people, what generally unforeseen outcomes may arise later, and so on. This wide-lens perspective is key in the dual use or biological and chemical security context, just as is the shorter focus on what is going on right there in your laboratory.

Remember that I am not proposing here that all scientific research has hostile dual use potential; rather, I am suggesting that you, as a practising scientist or associated professional working with scientists, need to be able to recognise potential dual use risks in the work you are associated with. In addition, this will help you to identify what impacts this may have on society on a wider scale. Hopefully, The Toolkit will help you to incorporate an awareness of dual use potential and its social impact into the daily overview of your work, just as you already incorporate biosafety and bioethics awareness into your daily routines.

Once biological and chemical security risk assessment is incorporated into your daily approach to your science work, you are likely to find that your notions of risk and potential adverse outcomes become much wider than hitherto. Don't worry unduly about this. People experienced the same widening of concerns when microbiology became understood in the 19th century and again when biosafety practices became a new norm in science and public health. Now that biosafety is embedded into our daily practice, we don't think it is unreasonable and we adapt our work to accommodate the requirements compliance places on us. We can do the same with biological and chemical security.

Chapter 3

Risks and Responses Fit
for the 21st Century

Definition of Terms

Let us look at some definitions first. I have referred so far to both 'biosecurity' and 'bio-chemical security'. These concepts sit alongside the concept of biosafety, which we are all already aware of, and compliant with, even if we do not all achieve it to perfection (see Chapter 8). The most common security-focused term used by various authorities to describe the response to the risks of biotechnology misuse to date has been 'biosecurity', which is often confused by non-scientists with 'biosafety'. However, as the increasing convergence of biology and chemistry becomes, more apparent in research and commercial developments globally, a more useful security-focused term with which to complement 'biosafety' may be 'bio-chemical security.' The Biological and Toxin Weapons Convention (BTWC) specifically includes *toxins*, even though these may be defined as chemical substances. Given also that the CWC covers toxins of biological origin, there is thus a significant overlap between the BTWC and the Chemical Weapons Convention (CWC).

This convergence needs to be recognised and considered in any discussion on biosecurity because it has practical implications within the laboratory as well as outside it. By utilising the term 'bio-chemical security', we may cover responses to the demands of both the BTWC and the CWC in a single term. It is arguably easier to use a short blanket term such as 'bio-chemical security' rather than the lengthy 'biological and chemical security' (you may not agree).

Having said all that, as many other authorities continue to use the term 'biosecurity' to encompass everything I have just outlined, please bear in mind that when I refer to 'bio-chemical security' this is simply an expansion of the meaning of the term 'biosecurity' used elsewhere. My definition simply 'covers the ground' to include security issues relating to chemical substances and processes as well as biological ones. If you would like to use 'bio-chemical security' as a means of spreading the idea, go right ahead. If you prefer to retain the term 'biosecurity', that is fine too. Just be consistent and make sure everyone knows exactly what your terms mean and cover. However we refer to these concepts, let us just bear in mind that they all come down to one thing — protecting valuable science from those who would wish to misuse it, either through 'our' accident or negligence, or through 'their' hostile intention.

Biosafety is itself an interesting issue. In the West, we take good biosafety practice for granted, wrongly. Anyone who has worked in a lab knows that despite all the obligatory training sessions, the posters on the wall showing 'best practice' and the ever-present threat of the biosafety officer appearing just as you are wiping up the coffee spill that dripped over your experiment, we do not adhere perfectly to all biosafety requirements on a daily basis. A quick look at a report of problems at the Center for Disease Control and Prevention in the USA[1] and equally worrying reports about major British labs[2] from as recently as 2014 should concentrate our minds (see Chapter 8). If we cannot assure ourselves of the robustness of our biosafety practices, what chance is there that our biosecurity practices will be any better?

---

[1] See Jocelyn Kaiser's article of July 2014. Available on: http://news.sciencemag.org/biology/2014/07/lab-incidents-lead-safety-crackdown-cdc. Accessed 27/12/15.
[2] See Ian Sample's report of December 2014 about British labs. Available on: http://www.theguardian.com/science/2014/dec/04/-sp-100-safety-breaches-uk-labs-potentially-deadly-diseases. Accessed 27/12/15.

In today's world of terrorism, unstable states and easy global travel, not to mention instant global communications, how can we be sure that our practice is really secure?

Amazingly, given this scenario, there is no agreed global definition of 'biosecurity'. The term is often mixed up with, or used interchangeably for *biosafety*. This is not helped by differences in language and semantics. The French word for biosafety is *biosecurité*, for example. This lack of clarity in definitions is a major problem. I have seen science dictionaries that define 'biosecurity' as 'biosafety'. Confusing! If we cannot agree, globally, on what we are talking about when we use certain words, then we cannot understand each other or agree meaningful ways forward. Worse, our efforts to mitigate bio-chemical security risks will be hampered or rendered useless.

We can only hope that at some point in the not too distant future, science will grasp this problem and do something to address it. We can all (usually) agree on the definition of *biosafety*, even if we do not all manage to 'do it' perfectly. Surely an agreed definition of *biosecurity*, or in this case, bio-chemical security, is not too hard to come up with? It ought not to be that difficult for the various formal science bodies to discuss and agree the meaning of such an important term. Given that most scientists would not even contemplate using inaccurate terminology in their daily practice, this does not seem to be an unreasonable hope.

Confusion, often enacted at a governmental level, can cause all sorts of issues above the level of the individual scientist. In the UK during the 2001 Foot and Mouth outbreak, much of the publicity referred to 'biosecurity' when what was referred to could have better been termed 'biosafety' (although biosecurity as I define it was an issue — what measures were in place to prevent the Foot and Mouth virus falling into the hands of ill-intentioned people? It would make a good biological weapon for anyone wanting to have a major adverse impact on the economy of a region or state). This was the case because in the UK the term commonly in use to describe responses to agricultural disease threats is 'biosecurity.' One could argue, of course, that as long as procedures are followed, it is not important how well terminology is understood, so does it matter if we talk about *biosafety* or *biosecurity?* Fair enough, to some degree. But in and around the laboratory, we need to *understand* in order to better *comply.* Thus the inadequacies of definition are a problem that is unlikely to go away until an international definition is agreed *and understood.* This is because the implications for non-compliance are potentially different: one focuses largely on just the laboratory and the other focuses on the laboratory *plus* everything and everyone *beyond the lab door.*

Convergence of Biosafety and Biosecurity

The World Health Organisation defines *biological* agents as those that 'depend for their effects on multiplication within the target organism, and are intended for use in war to cause disease or death in man, animals or plants' and '*Chemical* agents of warfare include all substances employed for their toxic effects on man, animals or plants.'[3]

It is obvious that there is a significant overlap in these definitions. My own definitions covering all of this are here:

- *Biosafety* is the containment of pathogens, toxins, and other dangerous materials through a range of risk assessment, risk mitigation, and performance-measuring activities;
- *Biosecurity/bio-chemical security* is the containment of pathogens, toxins, and other dangerous materials including chemical substances through a range of risk assessment, risk mitigation and performance-measuring activities **PLUS the containment of associated information, knowledge, data, equipment, and intellectual property through the same or similar activities.**

Do you agree with these definitions? If not, can you explain clearly to a colleague why you don't agree with them? How would you define biosecurity? Why? What is wrong with my definition here?

---

[3] World Health Organisation (1970) *Health aspects of chemical and biological weapons.* Report of a WHO Group of Consultants WHO: Geneva, Switzerland, p. 12. Available on: http://www.who.int/csr/delibepidemics/biochem1stenglish/en. Accessed 27/12/15.

We need to start and maintain an international dialogue around terminology. There is always room for improvement.

Biosafety is built on the theoretical ethical principle of 'do no harm' and is implemented as practical protection of the individual, the group and the environment, through containment, from unwanted exposure to harmful pathogens and other harmful agents. *Biosecurity*, or in our case looking at the demands of both prohibition treaties, *bio-chemical security*, is built on the same ethical principles of 'do no harm' and is implemented as practical protection through *adding* to biosafety the containment (or some form of appropriate restriction, whether temporary or permanent) where necessary, of information, knowledge, and communications that could result in either accidental, unintentional or deliberate harm arising from scientific and technological advances. Please note that I am *not* advocating the 'censorship' of science or saying that certain aspects of scientific endeavour must never be published. Rather, I am saying that when risk is identified in the carrying out and/ or publishing of certain work, then a formal framework for consideration for the risks — plus an effective response plan — need to be implemented in order to protect that science from misuse by others.

Biosafety stipulates that certain activities should not be undertaken, or undertaken only under strict regulation, due to the risk of exposure or escape of pathogens or other harmful substances from containers or from the laboratory. Certain risk-mitigating questions are asked prior to work commencing, and as work goes along; appropriate responses are then made in the form of containment processes and Standard Operating Procedures. This is all ethically-mediated even if it is not obvious. In the interests of *doing no harm*, biosafety imposes various limitations on personal and group *autonomy* in the pursuit of individual and group *safety*. *Consent* to this is required in order to work in the laboratory. *Benefit* to humankind is assumed.

Biosecurity likewise stipulates that certain activities should not be undertaken, or at least controlled or restricted, due to the risk of 'exposure' or 'escape' of information, knowledge and communications from the laboratory or office of the scientist or technician. Just as a pathogen can be dangerous, so can information, knowledge, and communications be dangerous if they provide a person with hostile intent with the necessary means to effect a biological or chemical attack. However, ethical *responsibility* is usually deemed to have ended once work is published or otherwise completed, as in biosafety. This is a crucial difference between biosafety and biosecurity. Biosecurity is mediated by the same ethical principles as biosafety. Unless practitioners are aware of the risks, and of a practical ethical framework within which to work, no ethical recognition can occur, and no risk mitigation can be put in place. In biosecurity, the scientist needs to look ahead, post-publication or post-research completion, and consider his/her responsibilities *as they can be reasonably foreseen at that time*.

Thus the ethics education of those working in science is itself an ethical issue. Effective, ethically-driven biorisk management (BRM) (combining both biosafety and biosecurity) is therefore one of the strongest tools in our global response to potential threats to humankind from biological and chemical weapons.[4]

Table 3.1 summarises the convergence of biosafety and bio-chemical security in theoretical and practical terms. Ethical consideration of this convergence is increasingly important due to the difficulties experienced by both scientists and the security community in dealing with biosafety and bio-chemical security as two separate entities. Indeed, treating biosafety and bio-chemical security as separate entities may actually lead to problems such as missing risks altogether, or characterising (determining the likelihood and consequences of a particular risk within a risk assessment) and evaluating (defining what level of risk is acceptable) these risks inadequately.

Have Scientists Always Worked for the Common Good?

The informed public is widely familiar with military examples of dual use, even if the term has not been used in the general media or by some other authors. I have already mentioned or considered the Manhattan

---

[4] See Sture, J. (2014) Dual Use. In *Global Encyclopedia of Global Bioethics* (Springer). Available on: http://www.springerreference.com/index/chapterdbid/398716. Accessed 30/12/15.

Table 3.1: The convergence of biosafety and biosecurity in theory and application

Biosafety	Bio-chemical security facilitated and supported by an ethics approach
Biocontainment Facilities	Physical and IT/digital security of premises, equipment, offices, computers, paperwork, etc., plus all other facilities — what ethical issues does this raise?
Biocontainment Practices	Physical, IT, digital and intellectual security of information, data collection, processing, analysis, storage and review of materials and other research activities; (e.g. Standard Operating Procedures, etc.) — what ethical issues does this raise?
Biocontainment Personal Protective Equipment (PPE)	Personnel competence and understanding, reliability, history and willingness to engage effectively with bio-chemical security — what ethical issues does this raise?
Decontamination & Disposal	Security of physical, IT and digital information, data/findings; security of deletion processes — what ethical issues does this raise?
Emergency Response & Biosecurity	A plan of what to do in case of a bio-chemical security breach — what ethical issues does this raise?
Shipment and Transfer of Biological Materials	Security of information, data/findings/processes etc., when being transferred between legitimate parties/places (e.g. insecure email and so on); approved lists of 'buyers' and 'sharers' — what ethical issues does this raise?
Strategies for Effective Biosafety Management & Communication	A robust plan to manage the containment or restricted dissemination of information, data/findings/processes — publication, conferences, secure discussions and so on (may decide that no restriction is needed) — what ethical issues does this raise?

Project (nuclear science used to build devastating bombs), the Australian Mousepox case (genetic engineering leading inadvertently to death or sterilisation in mammals) and Galston's fertilizer (beneficial plant science developed to use as a weapon). Clearly, the first of these examples was an intentional dual use of science, whereas the second and third examples were unplanned by the original scientists. They were still, however, adverse outcomes following on from an originally-reasonable scientific endeavour.

The overwhelming majority of people involved in the Manhattan Project had no idea of the nature of the overall 'project' (at its height in 1944, the Project employed over 122,000[5] people, although only a very small number were scientists). This is an ethical issue in itself, but as it was a wartime effort, who would have been prepared to question it? Ramshaw and Jackson became alarmed as soon as they noticed the unexpected outcomes of their mousepox research, but it continued and was published (the military apparently ignored their report). Galston had no idea, until it became obvious, that his work had been used by the military to develop a biological weapon to use in a foreign country. I have worked with colleagues who were previously involved in their native countries' active biological and chemical weapons programmes. At the time of their involvement, they did not all know exactly what they were involved in, although some of them suspected the nature of the 'bigger picture'. Even among those who did know what they were part of, what could they have done to *uninvolve* themselves? Everyone has responsibilities and needs paid work. Most of us need to support or contribute to a family budget. Most scientists want to build a career that will be successful, both for them and their discipline. In many countries, whistleblowing is not an option, and in those where it is institutionally enshrined in policy, the outcomes for whistleblowers are rarely positive.[6]

Dual use is clearly not a cut-and-dried issue in terms of responsibility. Nevertheless, in today's world of compensation, inquiries, human rights and accountability, we cannot afford to ignore the potential misuse of our work — it may come back to haunt us in ways that we find unbelievable right now. This is why it is so important to nip potential dual use, or manipulation of science for hostile

---

[5] *Manhattan District History* Book 1 — General, Volume 8 — Personnel. ('Manhattan District' referred to the US Army component of the Project). Available on: http://www.osti.gov/includes/opennet/includes/MED_scans/Book%20I%20-%20General%20-%20Volume%208%20-%20Personnel.pdf. Accessed 28/12/15.

[6] Martin, B. (2007), Whistleblowers: risks and skills, in B. Rappert and C. McLeish (eds), *A Web of Prevention: Biological Weapons, Life Sciences and the Governance of Research*, Earthscan, London, pp. 35–47.

ends, in the bud. Once it has started, it is hard to stop, with the potential 'fallout' affecting more and more people, many of them innocent bystanders in the workplace who have been drawn in unwittingly.

Until relatively recently, many scientists were in tacit agreement with the rest of the public in the belief that biological and chemical weapons were a thing of the past, or at least a long-ago relic of the First World War. Unfortunately, we still see reports of the use of chemical weapons in the news today. Chemical weapons were used by Iraq against Iranian forces and against the Kurds between 1983–1991[7] and there have been various reported incidents in the apparent ongoing use of these abhorrent weapons in the Syrian civil war, even after apparent decommissioning under the auspices of the OPCW in 2014.[8] There have been a number of peace-time bio-chemical attacks as well, as in the US anthrax letters of 2001.[9] This case has never been completely 'solved' to the full satisfaction of all parties, despite the common understanding that a named individual was responsible.[10] In Chapter 1 we also looked at the Aum Shinrikyo attacks in Japan, which were intentional and which were enabled by the active participation of scientists.

Do we really believe that any of these and similar attacks were enabled, supported and carried out *without* the involvement of

[7] BBC news website (2003) *Saddam's Iraq: Key Events.* Available on: http://news. bbc.co.uk/1/shared/spl/hi/middle_east/02/iraq_events/html/chemical_warfare. stm. Accessed 28/12/15; UN News Centre (2006) *UN releases report on Iraq's chemical weapons programme.* Available on: http://www.un.org/apps/news/story. asp?NewsID=18714&Cr=iraq&Cr1. Accessed 28/12/15.

[8] Organisation for the Prohibition of Chemical Weapons (OPCW) — United Nations Joint Mission (2014) *Thirteenth monthly report to the United Nations Security Council.* Available on: http://www.un.org/en/ga/search/view_doc.asp?symbol=S/2014/767. Accessed 28/12/15.

[9] Federal Bureau of Investigation. *Amerithrax or Anthrax Investigation* Available on: https://www.fbi.gov/about-us/history/famous-cases/anthrax-amerithrax/amerithrax-investigation. Accessed 28/12/15.

[10] Engelberg, S., Gordon, G., Gilmore, J. and Wiser, M. (2011) *New Evidence Adds Doubt to FBI's Case Against Anthrax Suspect.* Available on: http://www.propublica. org/article/new-evidence-disputes-case-against-bruce-e-ivins. Accessed 28/12/15.

scientists? Clearly not. So the question arises — who were the scientists involved in these attacks? Even if they did not engage in the deployment of these weapons, they played a role somehow in allowing or enabling the acquisition of the biological or chemical materials. Even if no scientists were involved in the actual attacks, where did the perpetrators get their science education (whatever that amounted to)? At what point(s) in the past did they acquire their skills and knowledge? Who is or was funding them? From where were they operating? From where did they acquire the necessary materials? Were they, or have they ever been, registered as members of a professional association? Did they have accredited scientific qualifications? From which institutions did they gain these? Are there any sanctions in place that could be applied to them? Did, or do, any of their colleagues have any idea about what they were/are 'up to' during their education, professional practice or post-education activities? Were there ever any suspicions or misgivings about them or their activities? Was anything reported or followed up? Can we afford to ignore the uncomfortable scenarios that these questions raise?

I am Insulted! My Work is Not a Danger to the Public

A significant problem for the advancement of bio-chemical security education and understanding lies in the fact that many scientists and science stakeholders are aggravated, understandably, by the sense of judgement and potential censure that goes with the acceptance of bio-chemical security as a daily risk inside and beyond the laboratory door. This aggravation is entirely understandable.

However, the situation needs to be understood from the perspective of the security community and the general public, who see the intentional or even the potential hostile use of biological and chemical technology as the *misuse* of that science. Obviously, the term 'misuse' expresses a judgement. This element of judgement, unfortunately, gives rise to further challenges between those working in security circles and those who work in the world of science and technology. Most scientists, already feel over-burdened with regulation and oversight. Many feel, with some justification, under attack by the 'biosecurity police.' To the security community, rightly or wrongly, scientists are often viewed as being unaccountable and almost a law unto themselves; this leads to scientists sometimes being viewed as a (potential) security threat.

Let us look at this for a moment, though. What other profession apart from lab-based life-science can claim academic and scientific freedom to effectively do what it wants simply because scientific advances enable new forms of work to be attempted? Is it possible for medical practitioners, pharmacists, optometrists, and so on, to do research or other forms of work within their disciplines — in other words, on humans or the materials of life — simply because they *can*? Of course not. We are all familiar with ethical concepts such as the Hippocratic Oath and most of us have some awareness of the various Christian, Jewish, and Muslim teachings and values (among others) that have historically informed healing and science practices and determined the inherent value of the human being and of life itself. Such theories and beliefs underpin our ethical approaches to the care of people as individuals and our duty to 'do no harm.'

I would suggest that one of the key differences between perspectives on ethics in the healing arts and sciences and those that seem, to the public, to pertain in science labs is simply this: the scientist who is a health professional examines and treats 'patients' — he or she is in direct daily contact and interaction with the patient and often the family and friends. We all get to know 'our' doctor, 'our' pharmacist, and so on. By contrast, the lab-based scientist is removed from this direct daily contact with the public and largely works out of sight in a privileged and 'secret' environment. We do not know 'our' research scientist. I would argue that if we, as lab-based scientists, were forced (not literally, I hope) to meet with Joe Public on a daily basis in order to explain and share our work, then we would start to think in a slightly different way.

I am not saying that scientists do not already consider ethics — we do, but this is generally done at a discipline-level and tends to be theoretical, while being almost totally subjective when it comes to practical application. This leads to most of us always saying that we are *definitely* ethical practitioners. I never yet met a scientist who thinks he or she is unethical. Yet these same colleagues are typically unable to explain what ethical framework they use to make decisions, what their philosophical position is, or what decision making processes they use to decide on ethical options (assuming that they recognise the ethical options in the first place).

This then leads to the argument in which a scientist may habitually say:

'I think doing *x* is acceptable. Therefore doing *x* is ethical.'

But this is a totally subjective process and decision. The problem is that the scientist in the country next door, or even in the lab next door, thinks:

'I think that doing *x* is not acceptable, so doing *x* is unethical.'

You are both convinced you are right, and both convinced that you are therefore 'ethical'. Hmmm. Back to the drawing board!

This difference in exposure to the opinion and reaction of the man and woman in the street on a daily basis is key to the perceptions of science held by the public, including the security community. Before you shout at me that you *do* work with the 'man and woman in the street' — your staff and colleagues — I would say this: your staff and colleagues are not quite the equivalent of the 'man and woman in the street.' They are at the very least people who are 'bought in' to science, either because they are scientists, or because their living depends on the success of scientists. This means that the views of those with whom you work are highly likely to be biased in favour of your typical scientific approach to ethics, just as colleagues are in any walk of life. This is where they may well differ from the 'general public.' Bear in mind too, that being associated with science, laboratories, scientific research, and so on is something that can bring a certain social status by association (even if not an economic one). Anyone who says that he or she works in a given scientific field, even in administration or allied areas, is often automatically accorded a certain respect that is not reflected in that given to other fields of employment. I am not saying that any of this is universal or right, but merely observing that it is another influence on opinion and belief. Factors such as these can be surprisingly influential in forming peoples' opinions of controversial issues.

For the rest of the world outside the laboratory, while lacking in a deep understanding of the scientific detail and the relevant theoretical, philosophical and practical issues, there is still an awareness of at least some of the risks and potential outcomes of the misuse of science. The public maintains an abhorrence of bio-chemical weapons and often simply sees scientists as arrogant people in labs who get up to dangerous things just because they can. So we have at least two sources of permanent potential criticism for scientists — the public and the security community.

None of this is welcome to us, operating under what we already consider to be a high burden of oversight and regulation. Nevertheless, this is the world in which we find ourselves. As I said earlier, scientists are the best-placed people to review their work for risk, and to devise appropriate mitigation activities to it to reduce or eliminate dual use risk. This is why it is vital that we scientists engage with security issues now, before governments and others impose it on us from a position of lesser knowledge and understanding.

Do we really want to see non-experts dictating what we can and cannot do in the laboratory? If not, then it is a good idea to take on board the worries of the security community, the public and civil authorities and engage with bio-chemical security ourselves. In this way, it will be possible to allay fears and retain the majority of our influence over our own work. It will enable us to recognise any reasonably foreseeable potential for misuse as we devise and carry out our work and also provide us with a skill set with which to review the work of others as we seek to prevent misuse in the future.

Acknowledging the Risks of Misuse

There are two typical responses that come up, in my experience, when the concept of bio-chemical security as a daily risk is mentioned to scientists and other technologists. Firstly, all the scientists I have ever worked with, anywhere, state unequivocally that 'My work is not a danger to anyone' and 'My work helps people'. Secondly, many go on to state 'Academic freedom! Scientific freedom! It is our right as scientists to do this work.'

These statements, made in good faith, illustrate a big problem. This is not a problem with scientists, or with the work they are doing. It is a problem of *recognising and understanding risk*. It is an entirely understandable problem. If you entered science to make a positive difference in the world, it is quite challenging to be told that your work could actually be used to make a negative difference in the world. If you have never been introduced to dual use issues before, why should you believe that it applies to you and your work? I have great respect for, and faith in, the scientists with whom I have worked. Their work does indeed help people. They have entered science as a profession in order to help others and to add to the common good, as well as to acquire personal benefits from science such as a job and an income. This is all good. However, while correct and good as far as it goes, this approach misses the point.

We have mentioned already the *intentional misuse* of science. This does not have to mean *your* intentional misuse of science. *You* need to protect *your* work from misuse by *others*. Misuse by others is still a misuse of your work, even if you know nothing about it (think again of Galston and his fertilizer). Further, while most reasonable people would agree on what is 'misuse' — applications leading to the death, injury or some disadvantage of others — even this is not without dissent. Who actually decides what *is* misuse?

If you are a religious fanatic intent on imposing your beliefs on those who do not currently agree with you, you would possibly state that using science to kill dissenters is not a misuse of science. This is the 'God's Will' argument that has underpinned crusades, jihad and burnings at the stake for centuries. Most readers will not recognise such people as 'reasonable' (I certainly do not), and take some comfort from the fact that misuse of science is only carried out by someone else, in another place and another country, with nothing to do with 'their' work. This is, unfortunately, the equivalent to the 'it will never happen to me' response that we all cling to when seeing disasters unfold on the evening news. Misuse of your work can be just as easily undertaken by people you know as by someone on the other side of the world. While we do not want to get to the point of vetting all staff and students for 'reliability' (although there are scenarios where this is done and is advisable), we should be looking at ways of mitigating risk in such a way as to avoid the more drastic security measures that nobody wants (yet).

Having said this, it is likely that, in the event of a single significant adverse event occurring and being traced back to a particular lab, unwelcome restrictions will follow as a means of allaying public fears and to boost security for the future. We only need to have one big adverse event — the misuse of a specific case of scientific research as a weapon either purposely or as the result of an accidental outcome — for the current 'it will never happen to me' perspective to change. Once the trail of breadcrumbs has led back to your lab, and to the work that you or your team did, which was possibly even funded by your government or a major funding body, then what will be your defence? To what will you be able to point to show that you took all possible and reasonable steps to recognise and mitigate the risks of misuse of your work? As the British Prime Minister Harold Macmillan allegedly once said, when asked what he feared most: 'Events, dear boy, events.' We all need to think about possible 'events' a lot more.

I recognise that I am using extreme scenarios here to prove a point. In practice, the overwhelming majority of scientists are, of course, reasonable, caring people who take their responsibilities seriously. They would not dream of purposely causing harm to others. That is good. But what about the scientists of Aum Shinrikyo? They must have sat in classes looking like good students at some point before they moved into terrorism. The scientists who worked on the Manhattan Project, and Galston, all trained as scientists originally without any idea that their expertise and work would one day be turned towards weaponry and war. So at what point do we start to look for potential dual use problems? Let me illustrate how easy it is for the weaponisation of science to occur.

A Hypothetical Example of Misuse Arising from Benignly-Intended Scientific Research

Let us say that you have been working on varying the size of aerosolised particles for asthma treatment delivery systems using inhalers. You want to improve the way in which certain substances can be delivered with a standard inhaler. That's all excellent stuff and has clear benefits to the public, the drugs companies developing the work, and you as a science researcher (you will probably win further grants when you are successful and get your name associated with a health-improvement benefit). The benefits are clear. You do not intend to harm anyone. You may even wish to avoid testing on animals. You are a very ethical scientist.

You work to recognised biosafety standards throughout your project. Your research has been funded by a major funding body and has attracted some positive reviews and even some media attention at conferences. A patients' group (The Asthma Association, patron: a leading politician) has taken an interest in your work and wishes to endorse your findings in due course if the research is successful. You respect ethical practice in science and your supervisors/boss signed your plans off as meeting ethical standards. You are determined to 'do no harm'. You write up notes as you go along, on paper and on a computer or tablet, as everyone does. You leave your lab coat in the lab. You always wash your hands according to SOPs and so on. You use the biosafety cabinets properly and all the rest of it. You carry out your experiments in accordance with all the correct and appropriate procedures in the laboratory. Your notes and findings are kept in notebooks, on a computer in the lab, on your personal laptop, on various flash drives in the pockets of several coats and jackets, and are often sent via email between your media devices. A lot of information is stored on your mobile phone. You write up your thesis or report and submit these via email and in print. A copy of your thesis or report, containing all the 'recipe' for your work, sits in the lab on a shelf with all the other recent work. A copy is in your supervisor's office on his shelves (so all his colleagues can see how many theses he has examined) and there is probably a copy in your institutional library. There may be a copy on an e-database for open access. You publish in full in journals and present it all at conferences. You and your supervisor even apply for a patent. Great, All good.

You have produced a fully-written up, clearly described recipe for how to produce a biological weapon. You not only did the work in full to develop and perfect this, but you wrote it up for others to read freely. Your published it. You trumpeted your success at conferences. You left your tablet, your phone and your flash drives — all of which contain vital information, around the lab, in the institutional café and even on the bus. You lost one flash drive altogether but did not worry as you had other back-ups. Even if your employer or supervisor had an embargo put on the work while patents are sought, the recipe is out there, or will be soon. Hmm. Did you do this on purpose?

Yes, in that you did the work for a good outcome — the better treatment of asthma sufferers. No, in that you did not plan on some ill-intentioned person taking your technique and applying it, not to the better dispersal of a drug via an inhaler, but the intentional dispersal of a dangerous biological or chemical agent via an inhaler. Mr or Ms Hostile can now use your work freely to fill an inhaler with something nasty (which could not have been delivered via an inhaler before you did your work and showed how to do it) and spread it all over the nearest airport by spraying it near an AC vent, or in the toilets, or over the check-in counter. Hundreds of people will be exposed to this agent unwittingly, before hopping onto assorted planes and taking it to their many destinations around the world. Job done. Are you responsible for this? If not, who is?

Is it acceptable for you to state that only those with the hostile intentions are to blame? Who is going to want to listen to that explanation when hundreds or perhaps thousands of people are sick or dying? Your rationalisation that you only did the underpinning work for beneficial purposes is probably going to look more like an excuse than an explanation to others. Not only will you have your boss on your back, but you are likely to also have your funders (who may never want to see a grant bid from you again), the editors of the journals in which you published the papers, the academic press, the government and possibly the security services on your case. The media will be knocking on their doors with great energy and they will naturally want to pass the buck to you. On top of all this, you will have the general public writing to their favourite newspapers, Twittering vigorously and maybe even picketing your lab. Your university department or commercial company suffers a big drop in applicants. Funding falls away. Lawyers for the families of the sick and dead may start hitting you with lawsuits. Oh dear. What price has 'scientific freedom' cost you now?

So what can you do in order to minimise the risks of this happening and to protect yourself and your work at the time of planning it and carrying it out?

Sources of Advice, Guidance, and Approval

For those working in science and carrying out research in higher education institutions, their university Ethics Committee would usually be a source of guidance, advice, and support. For those universities with actual Ethics Approval Panels, this system would be even more robust. Most commercial laboratories have access to some sort of ethics approval process and structure, even if it is in conjunction with another institution. Bear in mind, however, that at the time of writing this book (2015) there is still little recognition of dual use issues within life sciences, and therefore likely to be little awareness of them on Ethics Committees. However, given their very nature, most Ethics Committees could certainly contribute to support and offer help if asked. For US readers, you also have the NSABB to consult if needs be (see Chapter 2).

Your discussions with advisory boards and ethics panels or committees are important. Even if no dual use risks are identified in your work, the very fact that you are raising this issue will be instrumental in spreading awareness of dual use bio-chemical security issues beyond the laboratory and into other research areas where it may have an impact. On this basis alone it is worth you taking the time to introduce the topic to your own Board or Committee. Even if they have never heard of dual use, it is better for awareness to come from you now than from them later. If you can show how you have used the ethical tools outlined in this book (or others from elsewhere), you should be able to assure the members of any committee that you have 'grasped the nettle' in identifying and addressing the actual or potential dual use risks in your work. So raising these issues yourself need not be a risky proposition at the ethics committee. The other benefit is, of course, that if you as a reader are seeing these issues for the first time, you can introduce them to your Board or Committee yourself.

Another vital form of ethical support and guidance may be found through the work of professional associations. Most of these, particularly in the life science sector, have Codes of Ethics, consisting of clear ethics guidelines, or Codes of Practice, that members are required to observe or adhere to. However, to date there appears to be little recognition of dual use issues within many of these. You may wish to think about raising this with your own professional association.

I mentioned the BTWC and the CWC in earlier chapters. It would be time well spent to look into the information your country publishes in relation to complying with these treaties. Your country will have incorporated compliance into its national legislation somehow. See if you can find which legislation covers your work and then familiarise yourself with the relevant sections.

An Exemplar Threat Spectrum Assessment Tool

So what exactly is the range of threats that we could be looking at? Take a look at the following tables and consider how the examples and principles given may be related to your own work. You do not need to be working with anthrax, smallpox or other well-known potentially-dangerous agents in order to have security implications in your work. You may be working on something that seems beneficial and harmless, such as improving aerosolisation in medical treatments (see the hypothetical example given above), but have you considered how your advances could also be used as a weapon against the public if the technology is widely available? An example of exactly this mode of delivery was seen in the 1993 attack in Tokyo, when a liquid suspension of *Bacillus anthracis* was aerosolised from the roof of an eight-storey building by members of Aum Shinrikyo. The nature of the attack was not fully recognized or understood for some years, but has resulted in considerable attention being paid to security in Japan since then.

The first table below is adapted from one published by Graham Pearson, a UK expert in chemical and biological weapons who has worked in the field of science and security for decades. He has written extensively on the subject and offers clear and insightful opinion and expertise that provide an excellent oversight of the two prohibition conventions (the BTWC and the CWC). In his 2002 publication[11] he states:

'There is considerable relevance between the CWC and the BTWC for a number of reasons. First, there is a close relationship between chemical and biological weapons which is shown by the CBW spectrum...... This shows that the two Conventions — rightly overlap in the area of toxins as well as in the area of bioregulators and peptides with the CWC listing two toxins — ricin and saxitoxin in Schedule 1. Furthermore, both Conventions address dual-use materials and technology, both totally prohibit a class of weapons which cause death or harm to humans and animals primarily through inhalation or ingestion, and both have general purpose criteria in their basic prohibition.'[12]

Next, let us think again about some of the recommendations of the Fink Report (2004) and the Lemon–Relman Report (2006) that we looked at briefly in Chapter 2. The following Table 3.2 considers

---

[11] Pearson, G. (2002) *Relevant Scientific and Technological Developments for the First CWC Review Conference: The BTWC Review Conference Experience.* First CWC Review Conference Paper No 1. Available on: http://www.brad.ac.uk/acad/scwc/cwcrcp/cwcrcps.htm.

[12] For analysis of Biological Weapons, Available on: http://www.brad.ac.uk/acad/sbtwc/; for analysis of Chemical Weapons. Available on: http://www.brad.ac.uk/acad/scwc/. Accessed 28/12/15. (Type carefully — these two URLs are very similar, but different nonetheless). As a scientist, you need to be aware of the nature and scope of these Conventions, and this is an easily-accessible source of relevant information for you and your colleagues and students.

Table 3.2:   An exemplar threat spectrum[13]

Classical Chemical Weapons	Industrial Pharmaceutical Chemicals	Bioregulators And Peptides	Toxins	GM Biological Weapons	Traditional Biological Weapons
Cyanide Phosgene Mustard Nerve agents	Aerosols	Substance P Neurokinin A	Saxitoxin Ricin Botulinum Toxin	Modified/ Tailored Bacteria Viruses	Bacteria Viruses Rickettsia Anthrax Plague Tularemia

Chemical Weapons Convention

⬅————————————➡

Biological and Toxin Weapons Convention

⬅————————————————➡

⬅————————————➡

Poison

⬅————————➡

Infection

You could adapt this table for use in any scientific research area, for example, synthetic biology, nanotechnology, neuroscience, genetic modification, reproductive technologies, and so on.

some of the recommendations arising from these, along with prior work by Graham Pearson, who developed the table above.

Biorisk Management — Within the Lab and Beyond the Lab Door

Now that we have looked briefly at the scope of dual use risk, let us take a closer look at the nature of *risk* itself. Then we can look at some approaches to risk mitigation.

---

[13] Table adapted from Pearson, G. (2002) *Relevant Scientific and Technological Developments for the First CWC Review Conference: The BTWC Review Conference Experience.* First CWC Review Conference Paper No. 1 p. 5. Available on: http://www.brad.ac.uk/acad/scwc/cwcrcp/cwcrcps.htm. Accessed 30/12/15.

Table 3.3   An Exemplar Threat Spectrum: from Pearson through to Fink and Lemon–Relman and relates to a number of publications,[14] all of which are available online

Pearson	Chemical weapons.
Pearson	Classical biological weapons.
Fink 1	Experiments that would demonstrate how to render a vaccine ineffective.
Fink 2	Experiments that would confer resistance to therapeutically useful antibiotics or antiviral agents (confer resistance to pathogen invasion).
Fink 3	Experiments that would enhance the virulence of a pathogen or render a non-pathogen virulent.
Fink 4	Experiments that would increase the transmissibility of a pathogen.
Fink 5	Experiments that would alter the host range of a pathogen (target specificity).
Fink 6	Experiments that would enable the evasion of diagnostic/detection modalities.
Fink 7	Experiments that would enable the weaponisation of a biological agent or toxin.
Lemon–Relman	Recommends adopting a broader perspective on the 'threat spectrum.' This covers anything not included under Fink (e.g. DNA, Synthetic Biology and delivery technology).

Source: This table is adapted from an earlier version devised by Simon Whitby of the former Bradford Disarmament Research Centre.

Becoming accustomed to Bio-chemical Security in practice

Let us take a look at the potential sources of biorisk that scientists encounter regularly, whether knowingly or not. This will help to

---

[14] Pearson, G. (2002) *Relevant Scientific and Technological Developments for the First CWC Review Conference: The BTWC Review Conference Experience* First CWC Review Conference, Paper No. 1. Dept of Peace Studies, University of Bradford. Available on: http://www.brad.ac.uk/acad/scwc/cwcrcp/cwcrcp_1.pdf.; The Fink Committee Report (2004) *Biotechnology Research in an Age of Terrorism* (National Research Council). Available on: http://www.nap.edu/catalog.php?record_id=10827; The Lemon–Relman Committee Report (2006) *Globalization, Biosecurity, and The Future of the Life Sciences* (National Research Council). Recommendation 2 can be found on page 232. Available on: http://www.nap.edu/openbook.php?record_id=11567&page=1. Accessed 28/12/15.

identify where bio-chemical security problems may arise, or from where they may have already arisen.

We can view biosafety and bio-chemical security as two complementary responses to biorisk. These responses are implemented through *biorisk management*. In order to implement this successfully, we need to understand the terminology and the theory underpinning it. Most of us get confused over the difference between risk, hazard, and threat; we often use these terms interchangeably. However, in order to better grasp what we are working with and to clarify exactly what we are talking about to ourselves as well as to others, let us look at this in more detail.

Hazard, threat, and risk[15]

Many of us get these terms confused. Just as I attempted to define other terms in Chapter 2, I will try here to define these terms in as simple as way as possible.

Hazard

A hazard is a *source* — a thing, a condition or a set of circumstances, that has the *potential* to cause harm. It may be harmful only in certain settings and circumstances and not in others.

Using an example given in Chapter 1, the Australian Mousepox Experiment, hazards include:

- The recombinant mousepox virus that rendered the previously *non-lethal* mousepox virus *lethal.*
- Notes, results, reports, and other communications that may or may not reveal the outcomes, showing how the virus was altered.
- the *knowledge* of how to genetically manipulate a virus to render it lethal and overcome the beneficial effects of a previously given vaccine.

---

[15] I am grateful to Stefan Wagener of the Public Health Agency Canada for this conceptualisation of these terms.

It is worth noting again that a hazard may not be dangerous in every situation. For example, a set of notes giving out potentially dangerous information (that could be used for malicious purposes) is not a hazard (or is at least less of a hazard) if it is only exposed to reliable people and kept in a secure location. (Of course, how do we know who is 'reliable'? And if they are reliable now, will they always be so?)

As a domestic example, you do not consider a boiling kettle to be a hazard when you make your coffee, but you would consider it a hazard if your small child wanted to pick it up himself. There is a range of variables that make the 'hazard' dangerous. In the case of the kettle, you would want to look at the age of the person using it, their understanding of the dangers of hot or boiling water, the proximity of the kettle to children, whether it is switched on or not, whether or not it contains hot water, and so on. An empty kettle with nothing hot in it is not a scalding risk to a child, but it could still fall on a child's head if pulled off the counter.

This is, of course, a question of balance. At present, many in the security community believe that scientists must be prepared to curtail their usual dissemination methods in the case of 'dangerous' research. This resulted, for example, in the 60 day moratorium entered into by scientists working on the H5N1 virus in 2011.[16] The work was subsequently published (see Chapter 8) but this case in particular brought scientific practice into the bio-chemical security spotlight. Think also about what Ian Ramshaw said in 2008 about the mousepox experiment:

> Another issue was that you would never want to release a recombinant virus that you could not recover into the environment. No matter how many experiments you do to show that these viruses do not infect humans or other animals, there would not be sufficient clarity about the consequences of environmental release.[17]

[16] World Health Organisation (2012) *Report on technical consultation on H5N1 research issues*. Geneva: WHO. Available on: http://www.who.int/influenza/human_animal_interface/consensus_points/en/. Accessed 28/12/15.

[17] Weir, L. and Selgelid, M. J. (2010) The Mousepox experience: An interview with Ronald Jackson and Ian Ramshaw on dual-use research. *EMBO Reports* 11(1), 18–24. Available on: http://www.ncbi.nlm.nih.gov/pmc/articles/PMC2816623/.

How would this read if instead it referred to 'knowledge of how to achieve this' being released into the environment? Let us paraphrase this along these lines (please bear in mind that I am *not* putting words into the mouth of Ian Ramshaw; this is *my* paraphrase of his words to illustrate what I am suggesting is a hazard):

'You would never want to release *the knowledge of how to achieve the suppression of natural killer cells and total suppression of the adaptive immune response* into the environment. No matter how *good your methodology or how effective your safety measures and oversight procedures are*, there would not be sufficient clarity about the consequences of environmental release.'

Do you agree with this? If not, why not? If you do agree, how would you then revise your practice to accommodate this sort of hazard? Bear in mind that Ramshaw and Jackson's experiments had all been approved by the Gene Manipulation Advisory Committee (the predecessor of the Australian government's Office of the Gene Technology Regulator) and in the original application they indicated that there was a possibility that the virus could be highly immuno-suppressive, resulting in a lethal infection.[18] They just did not expect it to result in the deaths of previously vaccinated mice, which should have been immune to the effects of the virus. This work, and that of others, was part of the consideration of biosecurity undertaken in the 2004 Fink Report. The committee of authors highlighted seven types of experiment *'that will require review and discussion by informed members of the scientific and medical community before they are undertaken or, if carried out, before they are published in full detail.'*[19]

Experiments such as the mousepox work came under these two recommendations:

1. Would demonstrate how to render a vaccine ineffective. This would apply to both human and animal vaccines.

---

[18] *Ibid*, p. 19.

[19] Committee on Research Standards and Practices to Prevent the Destructive Application of Biotechnology (2004) *Biotechnology Research in an Age of Terrorism*, National Research Council. Washington DC: National Academies Press, pp. 5, 114.

2. **Would enhance the virulence of a pathogen or render a non-pathogen virulent.** This would apply to plant, animal, and human pathogens.

Threat

A threat is *a person* rather than a thing, a condition or a set of circumstances, who has the potential or intent to cause harm to someone or something. This could be anyone in any situation. Notice that a person can be a potential threat or an actual threat, depending on the context. To some extent, we are all a threat if we consider the possibility of dangerous mistakes. However, taking a more pragmatic view is required if we are going to do any work at all. A person may not be an actual threat — may not have ill intent at one point in time, but may change his views and become an actual threat at another point in time, making use of what he learned earlier, when he had no ill intent.

In another scenario, a person could be a potential threat *inadvertently* through negligence or lack of training or understanding of the risks in what he is dealing with. This could result in an accident or a failure in safety/security due to *simply not knowing the risks.* This could be the student who is left unsupervised with a challenging experiment, or a contract researcher who has not been trained properly in the use of a specific piece of equipment, for example. It could even be a maintenance engineer who is lax about the responsibilities of his job who misses a fault in the autoclave that users may not notice. Complicated!

Clearly we cannot recognise or contain all threats at all times. But we are now advised that as scientists we *should be aware* of the potential threats to our work — people who may wish to use our work for hostile purposes as well as the possibility of people causing adverse outcomes by accident or neglect, or through lack of knowledge. Is this possible for us to achieve? I would say that it is not possible to achieve this to perfection, but that we can minimise the risks by at least doing what we *can* to take care of our work in such a way as to lessen the likelihood of our work being misused or the subject of adverse outcomes resulting from negligence or ignorance. The point here is, of course, that we have to actually *recognise* bio-chemical security risk in our practice, in order to be able to counter it.

Risk

Risk is facilitated by either a hazard, or a threat, or a hazard and a threat, in any one situation. Multiple hazards and threats may be present (think of an entire class of first year undergrads let loose in a lab). The hazards and threats may be recognised, or not.

The *severity* of the risk is influenced by the *likelihood* of the adverse event occurring, and the *seriousness of the consequences* arising from the event if it occurs. Therefore we may describe risk as the combination of the *likelihood* of a hazard or threat resulting in an adverse event PLUS the *consequences* of that event. *Likelihood* is the probability of an event occurring. *Consequence* is the severity of an event.

Remember that a risk can arise from the actions of a hazard or a threat, or a hazard *and* a threat. It can arise intentionally or unintentionally. Even if it is not recognised, it may still be present. One of the risks of the mousepox experiment was that the findings could be used to make a biological weapon using the knowledge that a virus could be made lethal and overcome the protective effects of previously given vaccines.

The *likelihood* of this happening depends on what measures scientists take to prevent this knowledge falling into the 'wrong hands' and the potential for misuse that arises once the research is 'out there' (notes lost, emails intercepted or copied to the wrong people, papers published, research presented at a conference, reported in the media, and so on).

The *consequences* of this knowledge being implemented in a hostile manner could be huge (many deaths, many people ill and many people in fear, with potential political fallout, civil unrest, economic impact and so on resulting). One possible outcome could be the use of this science as an actual weapon of war.

Many readers will now be thinking, 'Well the mousepox study was published in 2001 and nothing bad has happened yet.' But does this

RISK may be conceptualised like this:

BEFORE — Factors that increase or decrease the chance of the event occurring (likelihood)

Adverse
Event

AFTER — Factors that affect the nature and scope (severity) of the consequences

Figure 3.1:   Definition of risk.

mean that we should carry on with this sort of research without taking extra care to prevent such risks occurring now and in the future? Would you want your experimental work to be exposed as the source of a major security incident in the future? How would you be able to show that you had taken all reasonable steps at the time of the research to act responsibly with this sort of risk in mind? At least if you can point to all the mitigating actions you took at the time, you would have a level of personal and professional indemnity against the public backlash which would undoubtedly follow.

As well as thinking about the future, can you be sure that you have not inadvertently worked on dual use-potential projects already? As Ian Ramshaw said in 2008:

> I thought about this and realized there is another dual-use dilemma — and one that has not received so much attention. We created a transmissible virus that does not kill the individual but makes them sterile. That is as bad as making a virus that kills the individual. The principles were shown for mice; the principles were shown for rabbits; and there is no reason to think that similar principles would not apply to humans. I am only just realizing now that even before the so-called mousepox IL-4 experiments, we were already undertaking 'dual-use' experiments.

Table 3.3:   A Bio-chemical security risk spectrum (after Taylor, 2006)[20]

Naturally occurring diseases (hazards)	Re-emerging infectious diseases (hazards)	Unintended consequences of research (hazards and threats)	Laboratory accidents (people — therefore threats)	Lack of awareness (people — therefore threats)	Negligence (people — therefore threats)	Deliberate misuse (through people — therefore threats)
Natural environmental exposures, therefore a risk	Can lead to environmental exposures, therefore a risk					Devised to produce environmental exposures, therefore a risk

Unintended or accidental	————————————————⟶	Accidental or intended

Can you look back on all of your work and satisfy yourself that there is no dual use risk to any of it?

What Sort of Risks are We Faced With?

Table 3.3 illustrates what may be called a Biosecurity Risk Spectrum. While this focuses on biological diseases and hazards, do remember that chemical and toxin hazards come under the same scope of risk. The table identifies a range of hazards (things) and threats (people, either through negligence, accident or ill intent) that could lead to unwanted consequences through posing a risk. You can easily see that the hazards on the left of the table are 'naturally' occurring in the environment, whereas those on the right are the result of negligence or are planned in the presence of certain conditions that enable them to happen. We should remember as well that the nature and source of the hazard or threat have an impact on the way we address the risk of unwanted outcomes. If your boss is actually a threat through his slack attitude to some biosafety procedures, what could you do about it realistically? This is where a collegial approach to bio-chemical security can help to implement and support new (or old) ideas in practice

---

[20] Taylor, T. (2006) Safeguarding advances in the life sciences. *EMBO Reports* (European Molecular Biology Organization) 7, Special Issue. Available on: http://onlinelibrary.wiley.com/doi/10.1038/sj.embor.7400725/abstract.   Accessed 28/12/15.

in any facility. If there is a safe forum in which queries and concerns may be raised, this makes it easier to tackle the more difficult issues we all face in our professional relationships.

You may be wondering how accidental bio-chemical security risks can occur. This is actually quite simple. Most of these examples relate to biosafety issues that we take for granted in the West, but can still be the source of problems. For example:

- careless processing or logging of samples can result in the negligent 'loss' of materials; the missing materials may not be noticed for some time, if at all,
- leaving notes and records around, containing sensitive information about potentially dangerous processes, could allow information to fall into the hands of those who may not use it ethically,
- laboratory accidents can allow the taking of 'samples' by those not authorised to do so,
- inadequate or misleading labelling can enable the removal or mix-up of materials, intentionally or accidentally,
- failure to follow standard operating procedures, or the failure to maintain equipment according to the manufacturer's instructions, can lead to biorisk situations,
- inadequate disposal of contaminated materials can pose a hazard and allow the acquisition of dangerous materials by those who wish to misuse them,
- inadequate infrastructure — for example, are your waste disposal systems appropriate and well maintained? Do you only carry out the level of work that is allowed in your lab? Is there appropriate and suitable training available to all staff who work in your facility? How can you be satisfied that once waste matter has left your premises, it is properly disposed of?
- and so on.... see Chapter 8 for real-life examples of some of these problems.

So what can we do to reduce the risk of unwanted outcomes from our benign research and scientific practice?

Biorisk Management in Practice

A good place to start with this is the WHO's 2006 document *Biorisk management: Laboratory biosecurity guidance.*[21] Most readers will be familiar with this document. Here, we can see that BRM strategies are devised and implemented in order to:

- Reduce the risks of unintended exposure to pathogens and toxins;
- Minimise the risk of loss, theft, misuse or diversion of these to hostile purposes to acceptable levels;
- Provide reassurance and assurance that effective measures are in place to achieve these ends;
- Provide a framework that supports and promotes effective biosafety, biosecurity, ethical behaviour and appropriate training in the institution.

Another familiar key document to use in preparing BRM activities is the CWA 15793 (2008, 2011)[22] document of the European Committee for Standardization (CEN), which was supplemented by a further document CWA 16393(2012).[23] The second document provides guidance on how to implement the concepts described in the first.

We can look at the entirety of BRM under three banners.[24]

---

[21] WHO (2006) *Biorisk management: Laboratory biosecurity guidance.* Available on: http://www.who.int/csr/resources/publications/biosafety/WHO_CDS_EPR_2006_6.pdf. Accessed 28/12/15.

[22] CEN CWA 15793 is available from a range of websites but can be accessed here: ftp://ftp.cenorm.be/CEN/Sectors/TCandWorkshops/Workshops/CWA15793_September2011.pdf. Accessed 28/12/15.

[23] CEN CWA 16393 is available on a number of websites if you search online.

[24] This BRM framework is adapted from scheme taught by Sandia Laboratories, New Mexico.

- **Identifying and responding to the risks**
  - o identify what the risks are in your facility (physical, personnel, intellectual property),
  - o characterising risks (the process of determining the *likelihood* and *consequences* of a particular risk within a *Risk Assessment*),
  - o evaluating risks (the process of determining, subjectively, whether a risk is *high* or *low*, and whether it's *acceptable* or not).
- **Carrying out risk-mediating practices**
  - o identification of appropriate and suitable actions to minimise or eliminate identified risks,
  - o development and implementation of control measures to reduce or eliminate the identified risks.
- **Building in regular reviews**, in terms of:
  - o how well identification of risk and mitigation measures are carried out,
  - o personnel performance reviews and checks,
  - o equipment performance reviews and checks,
  - o process performance reviews and checks,
  - o appropriate revision actions to tackle any new or recurring risks that checks or ordinary practice identify.

A range of control processes and procedures are implemented to achieve best practice in relation to all of this.

This is not the place to go into detailed discussion of all the aspects of BRM. These bullet point lists are included here simply so you can see the kinds of issues at stake in bio-chemical security. Most of you will be familiar already with the issues around biosafety and BRM in the lab; this is just adding a further strand to consider.

I can understand those of you reading this who want to shout 'Oh no! Not more processes and regulation!' I get it. But all I am saying is that once you have become familiar with bio-chemical security you will be able to deal with it just as quickly as you do with biosafety. Nothing being proposed here is difficult or impossible to achieve. Bio-chemical security could be 'taken on' by existing Biosafety Officers. If preferred, a colleague could take on Bio-chemical Security in collaboration with the Biosafety Officer — although I would suggest recruiting people to these tasks who get along well, in order to avoid clashes of opinion and failure to align ideas, policies and practices. Besides, in terms of reviews, sharing between the biosafety and the bio-chemical security strands of BRM is vital — so this needs to be set up in a way that will work in practice.

Hopefully, this chapter will have highlighted for you some of the main concerns around your day to day work in and around the lab. The next chapter focuses on ethics as a topic and shows how you can begin to assess your existing activities under various ethical principles.

Chapter 4

Ethics and You:
What Does it All Mean?

Ethics, Values, Morals — What's the Difference?

In all the years I have been working in higher education I have always been amazed at the general attitude towards the subject of ethics. Without exception, all the students and staff that I have worked with have, prior to ethics training, considered ethics to be a subject that they do not need to be taught or to learn. Everyone believes that they automatically 'know' ethics already. I have often wondered when this seemingly miraculous transfer of knowledge occurs — are we born with it? Do we somehow *absorb* it from somewhere? *How* do we already *know* ethics? How do we know that we know ethics?

What people are doing here is confusing their personal values with public ethics.[1] I believe that this is what is causing the often

---

[1] Sture, J. (2013) Moral Development and Ethical Decision-Making, in B. Rappert and M. Selgelid (eds) *On the Dual Uses of Science and Ethics: Principles, Practices, and Prospects.* Canberra: ANU E-Press, pp. 97–120. Available on: http://press.anu.edu.au/?s=On+the+Dual+Uses+of+Science+and+Ethics. Accessed 28/12/15; and Sture, J. (2010) Educating Scientists about Biosecurity: Lessons from Medicine and Business in B. Rappert (ed) *Education and Ethics in the Life Sciences: Strengthening the Prohibition of Biological Weapons.* Canberra: ANU E-Press. Available on: http://press.anu.edu.au/titles/centre-for-applied-philosophy-and-public-ethics-cappe/education-and-ethics-in-the-life-sciences/. Accessed 28/12/15.

antagonistic attitude towards ethics that we see in the media, in our places of employment and underlying many compensation cases. Most people see ethics training as an insult to their personal values — as if the 'ethics police' are somehow saying that they need to be taught what values they should hold — as if they do not already have any. This is not the motivation of ethics tutors, but the fact that this view is common is, in my opinion, down to the poor communication methods of many of the 'ethics police'. In my experience, once a good discussion gets going in a class or group around ethical frameworks, people get really into it and love it. Even those who are challenged by it — honestly. Having said that, there is always one (may be two) who consistently fights ethics. Good luck friend, I will wait to see you on the front pages one day.

We all approach ethics from the perspective of our own set of private values. Some people also refer to 'standards'. When I say 'values', I am referring to the standards of behaviour, the understanding of right and wrong and the principles guiding our lives that we learned from our family first and our community later. We all think that our own values are the 'right' values. We may tend to use the term 'morals' instead of 'values'. This is where a lot of confusion lies. It may be useful for you to stop here for a moment and see if you can work out the difference between 'values' and 'morals'. I have never yet managed to achieve this effectively.

One simple way of looking at this is to divide up our ideals, or our attitudes to right and wrong, this way:

- I have a set of private, personal values that I grew up with (family influences) and adapted as I moved into different circles (school, university, work, social life, etc.);
- I have a set of public ethics that I hold to and use in the non-private arena (outside my home), even if I do not personally agree with them. This set of public ethics may be driven by social pressures such as the need to get and keep a job, to gain advancement and to be seen as 'acceptable' in public.

The two sets of ideals (private values and public ethics) *may* be a complete match, or they may differ in some aspects; where they differ, we need to formulate a way of managing any tensions this may cause.[2]

In teaching biosecurity classes and workshops in Western countries, I have always tried to avoid using the term 'morals' simply because in many Westerners' eyes the term is often loaded with the unwanted baggage of religion (I do not think that is necessarily true, but in order to avoid religious clashes in class, I tend to refer instead to 'values' — feel free to disagree with me). I want, generally, to keep religion *out* of the discussion around ethics in the West and to concentrate instead on looking at 'ethics' as a set of universal values that we can all sign up to, regardless of our religious views (including atheistic views). This is, of course, ironic, given that much of ethics has come down to us in the West through religious frameworks, especially the Judeo–Christian traditions. Unfortunately, in my experience, most Western scientists want to be seen as 'non-religious' — certainly in public at least. Those who are happy to advertise their religious views are often criticised, and in order to be taken seriously, seem to be required to dance some sort of tightrope between 'rationality' and being deemed by their colleagues to believe in the equivalent of fairies at the bottom of the garden.[3]

However, just to contradict myself, I *have* successfully used religious values as teaching opportunities when I have been teaching or facilitating biosecurity meetings or classes in the Middle East, North Africa or the Asia-Pacific region. This has also been useful occasionally in former Soviet states where orthodox Christian values still hold an appeal. In the Muslim world, this approach has been very

---

[2] Sture, J. (2013) *ibid.*

[3] See anything by Richard Dawkins, plus, for example: Gervais, W. and Norenzayan, A. (2012) Analytic thinking promotes religious disbelief. *Science* 336(6080), 493–496; balanced somewhat by McLain, S. (2013) It's a big, fat myth that all scientists are religion-hating atheists, *The Guardian* 04/03/13, Available on: http://www.theguardian.com/science/occams-corner/2013/mar/04/myth-scientists-religion-hating-atheists. Accessed 24/11/15.

successful because the general population in such countries is far more open to consider religious values as being *useful as tools for security*. I realise that to many Westerners this is incredible and almost unbelievable, given what we see on the evening news most days — but there it is. The only place I have ever had negative feedback about incorporating religious ideals into bio-chemical security was in Washington DC, where I described using class-supplied Muslim ideals to support bio-chemical security; I was told in no uncertain terms that I was building 'Islamic biosecurity' as if this were a bad thing. Clearly the US audience did not wish to incorporate Muslim ideals, which they considered inherently dangerous, into any issue of security.[4] Needless to say, some Western science-oriented audiences are happy to promote their own religious ideals when it suits,[5] with religious influence frequently being cited in decisions around abortion, embryonic stem cell research, assisted reproduction and euthanasia, to name just a few topics. Presumably it is acceptable to utilise your own religious ideals if it suits your arguments at the time.

In northern Africa, in Jordan and in Iraq, I have encountered only helpful input from Muslim (practising or not) scientists who are keen to use their religious ideals to support good bio-chemical security practice. We all need to be more aware of this and bear it in mind when considering our ethical responsibilities as scientists. Personally, I do not mind what religious value-set (if any) you want to use as long as it considers all people as equal and worthy of protection from harm and is not of an extremist nature (but who defines 'extremist'?). This means

---

[4] If you would like to read a little more about this situation, take a look at pp. 26–27 and elsewhere in my report: Sture, J. (2013) *International Biosecurity: Engagement between American and MENA Scientists* (a commissioned consultancy paper) as part of a Project on Advanced Systems and Concepts for Countering Weapons of Mass Destruction (PASCC), Center on Contemporary Conflict, Naval Postgraduate School. DOI: 10.13140/2.1.5043.2328. Available on: http://www.worldscientific. com/worldscibooks/10.1142/q0027#t=suppl

[5] See Western religious influences on start-and-end of life debates, and for example: Herman, C. (2014) Study: 2 Million U.S. Scientists Identify As Evangelical. *Christianity Today*. Available on: http://www.christianitytoday.com/ct/2014/february-web-only/ study-2-million-scientists-identify-as-evangelical.html. Accessed 24/11/15.

that men and women must be considered equally important, along with the many cultures, nationalities, ethnicities, religious groups, political groups, and so on that exist in the world. In public ethics we are talking about recognising and implementing standards of right and wrong that must treat all people as equally valuable and to whom equal protection from harm is due. If appealing to your religious values helps you to do this, then go ahead. If you prefer a secularist approach, follow that. As long as your views result in respect for others' rights and a commitment to protection for all, that's probably the best we can aim for.

Ethics and 'My' Values

If you ask your friends or colleagues 'Are you ethical?', they are almost bound to reply 'Yes' even if they have no clue what you are talking about. Who is going to voluntarily declare themselves 'unethical'? The problem here arises because we all tend to believe that everything *we* think is *right*. In our minds, it follows, therefore, that this must be automatically 'ethical.' Another way of saying this is demonstrated here:

> If I think this is right, then it is ethical — because *I* think it is right;
> I am an ethical person, therefore what I think is ethical.

However, as we saw in Chapter 3, this is a highly subjective decision-making process. It is also a logical fallacy.[6] You might think that washing your hands 'this way' is correct, and therefore 'ethical,' while your colleague next door thinks that doing handwashing his way is better — and is therefore 'more ethical' than your way. You may both be washing your hands incorrectly of course — in which case you are both acting 'unethically' in the lab.

There is a balance to be struck between our private values and our public ethics. Key to achieving this is the recognition that what *we* think is right may be seen as *wrong* by someone else. Both of us think

---

[6] If you want know more about logical fallacies. Available on: https://yourlogicalfallacyis. com/. Accessed 23/11/15.

our view is the 'normal' view. In order to get along socially and economically, sooner or later we have to start compromising on what we think is right and wrong. This does not mean that we need to abandon our private values, but that we need to compromise on them in some contexts to achieve a desired end.

Let's think for a moment about how we picked up our private values. For most of us, our values or our morals have been handed down to us by our families, shared between friends and reinforced by the society in which we grew up. Typically, our values are centred around the role, rights and responsibilities of the individual, the family, and the community. Relationships between men and women, the place of children, and between the home and the public sphere are all mediated by our cultural backgrounds.

Once we have entered nursery or school, our values are then expanded to accommodate the non-home world and public life. These values have been further developed by our own life experiences, either good or bad. Our early views of our family's values, or those of our culture, may have been reinforced by our experience, or our values may have been abandoned or adapted as we grew up. Whatever or however our values have changed over the years, we have all been challenged by the need to accept that others' values also need to be recognised and accommodated to some extent in order for society to remain peaceful. The problems arise when someone does not accept the values of others and tries to impose his own values onto others. The 'others' may not wish to accept these values. Trouble ensues.

In the workplace, in our case in the science class and the working laboratory context, we may be confronted with behaviours and activities that clash in some way with our private values. This is the intersection at which our private values must usually give way to our public ethics if we are to be able to work effectively in the public sphere. To use a blunt example — you will not want to be in charge of an abortion clinic if you are a practising Catholic. I could give many other examples here, but you get the idea. For most of us, the clash of values we typically experience is not as emotionally draining and painful as the Catholic being forced to work in the abortion clinic. However, we need to learn to recognise when clashes occur in order to be able to address them. Sometimes they are subtle, but even subtle clashes of values can cause problems.

Dilution of 'My' Values and High Standards at Work and in Class

One problem that appears in all educational and work contexts is that of hierarchies. In any school, university, lab, office or conference, there are invisible as well as visible hierarchies in place. Unfortunately, there is plenty of research available showing that ethics are often compromised by the effects of such hierarchies. Let me illustrate.

I mentioned hand-washing as an ethical practice in a section above. Before you start wondering why I am making an ethical point of such a mundane issue, read on. I used to work in the operating theatres (ORs if you are from the US) of a major English teaching hospital. An experience on my first day in theatre provided a perfect example of how ethics typically operates in the workplace where hierarchies are entrenched. As theatre staff, we were trained to 'scrub up' according to strict procedures, and before each operation we would line up at the scrub sinks and scrub up to our elbows for the prescribed length of time using soap and then an iodine-based wash (from memory I think it was five minutes with the iodine wash after 2 or 3 minutes with soap first) prior to putting on our surgical gloves and gowns. The surgeon involved was a senior professor who held a Chair of Surgery at our local university. He would always tell us, while we were standing at the scrub sinks, how we ought to scrub up for double the prescribed time, as in his opinion, the 5 minutes section of the scrub-up with iodine should be 10 minutes. Our tutor told us to ignore him and scrub for the 5 minutes, which we did. Each operation was thus delayed by an extra 5 minutes while The Prof did his extra scrubbing. Great, you might think. What an ethically-concerned surgeon. He wants to do it all to the best of his ability. How safe his patients must be. How admirable. Which it would have been, had he not ended his 10 minute scrub every time by turning off the taps with both of his hands. I asked our tutor, on the first day when I observed this, why nobody told him not to turn off the taps with his hands after scrubbing up. She replied 'Because it is not our place to do that.' Ethics, anyone?

In this instance, we see a perfect picture of how ethical standards become diluted for reasons of power relationships. This case is typical of what goes on in 'real life'. If you want to read a little more on how hierarchies and power relationships can dilute the initially-high ethical standards of medical students and business managers, take a look at a chapter I wrote in 2010.[7]

The reason I am providing this example is simply to show how difficult it is in practice to implement ethics effectively when the people 'at the top,' or the people with either the perceived power or actual power, do not act ethically. This is why it is vital that ethics and sound ethical practice needs to be taught from high school up, and continued post-education in Continuing Professional Development (CPD) activities. Unless everyone is 'bought in' to sound ethical practice, all of our efforts will be hampered.

In the following sections I will show how we may be able to effectively compromise our private values and personal ideas in the interests of an effective public ethics of bio-chemical security. I do not have all the answers, but I do have some ways of moving towards them. It will be helpful to remember that you are already 'doing' ethics when you engage effectively with biosafety — and that is not usually too painful. So applying the same compromise to achieve effective standards of bio-chemical security practice is not too painful either.

Ethics as Compromise of Your 'Rights' and Values

A recognition of the existing role of ethics in science gives us an opportunity to identify and 'sign up to' an agreed set of values that

---

[7] Sture, J. (2010) Educating Scientists about Biosecurity: Lessons from Medicine and Business, Chapter 2 in B. Rappert (ed) *Education and Ethics in the Life Sciences: Strengthening the Prohibition of Biological Weapons*. Canberra: ANU E-Press. Available on: http://press.anu.edu.au/titles/centre-for-applied-philosophy-and-public-ethics-cappe/education-and-ethics-in-the-life-sciences/.

will enable us all to get along and work to the good of society. That's the aim, at any rate. Ethics is often *compromise*. That is no bad thing if it allows us to avoid or minimise risk and tensions.

My favourite definition of ethics is this:

"Ethics is a matter of principled sensitivity to the rights of others"[8]

I like this definition (there are many others) because it implies some useful concepts.

- 'principled sensitivity' requires
  - o Principles underpinning actions and thoughts — some sort of codified, agreed or assumed framework within which to operate.
  - o Sensitivity (being aware of difference and being prepared to accommodate it to some agreed, suitable degree — or not) as an approach with which to look at other people and as a means of implementing ethical principles.
- 'rights' highlights entitlements of
  - o The right to life, autonomy, privacy, respect, information, not to be put in harm's way and so on.
- 'others' refers to
  - o your family, friends, neighbours, colleagues, people you pass in the street, people you don't know — everyone *who is not you*.

No matter how many definitions of ethics I come across, I always come back to this one because it is such a useful way to look at the *application* of ethics to *people*. As this book is all about applied ethics, and I am talking about the application of ethics *to other people* in the context of safe science, whether directly or indirectly, it seems like a useful definition to use. If you have a better definition, then use that. Just remember that it has to be *applied* and it has to accommodate the rights of *other people*.

---

[8] Gilbert, N. (2001) *Researching Social Life*. London: Sage, p. 45.

Different Approaches to Ethics

Without going into too much background detail, it will be helpful to first look at some of the various forms of ethics before we go any further. This is simply to briefly explain one or two approaches that you may have heard of before we move on to the practicalities of 'doing' ethics in support of the security of our work. It is not an exhaustive or complete list — just a very brief introduction to some forms of ethics.

Applied Ethics — the approach of The Ethics Toolkit

Remember that we are concerned in bio-chemical security in science with *applied ethics*. This is simply the putting of a theory into practice — taking it from the page and the philosophical ivory tower and *doing* it in real life. The question is, how do we do this?

In order to *apply* ethics to the real world, we have to 'operationalise' the theory of ethics in order to put it into actual *practice*. The term 'operationalisation' comes from social science, but do not let that put you off. We need to come up with a framework that we can translate from the page to real life. In other words, we need to find a way to incorporate the theoretical demands of ethics into daily practical use.

We need to consider what ethics actually is (feel free to think of 'ethics' as a singular or as a plural noun — there is no clear agreement, but I use it as a singular noun). How do we know that someone is 'being ethical'? I was once in a class with some Management PhD students, and their version of ethics (a tongue-in-cheek version, I hope) was 'we promise never to give out more in bungs than we pocket in back-handers'. You get the drift. 'Being good' and 'following the rules' are not sufficient as proof of ethical behaviour. The rules may be unethical themselves.

To illustrate operationalisation, let us think about the theoretical concept of social class or socio-economic status (whether you agree that it exists or not). I am using this because it is a fairly easy example to grasp, even if you are not British! How might we define what 'class' or socio-economic group a person belongs to? One simple way would be to find out what daily newspaper he reads. This is not a fool-proof method of course. The Queen may read a tabloid newspaper ('popular' papers — in the UK these include The Sun, the Daily Mail, the Daily Mirror, and so on) for all I know, but we would probably assume that she reads the 'broadsheets' by choice (The Times, the Telegraph, the Guardian, and so on) — apologies to non-UK readers, but you can substitute your own newspapers here. Another way of identifying 'class' would be to look at where a person was educated — private school or state school. We could look at post codes (addresses tend to be socio-economically dependent) or at which supermarket a person buys his groceries from. What sort of car does he drive? What has he named his children? What are his favourite TV shows? And so on. None of these operationalisations are fool-proof, or definitive, but taken as a group at a population level, they can arguably offer some degree of insight into socio-economic grouping. The key here is that we have identified a group of behaviours or characteristics that enable us to come to some understanding of what social 'class' a person may belong to. In the science context, we can look at certain behaviours and characteristics of scientists that may show us where they stand 'ethically'. Remember, of course, that *doing* something does not mean that you *agree* with it, so you can't assume a value from what you see someone doing or saying.

In applied ethics, everything is dependent on what the host culture (within which law and custom are developed and codified) values legally, socially, economically, and morally (yes, religious belief is important here, even if only as a historical influence). In the West, the most important ethical issues to consider in the research field are derived from a range of sources but focus almost totally on the *rights* of the individual. They are complemented by the *responsibilities* of the researcher or scientist. It is important to recognise that rights and responsibilities are very strongly mediated by cultural values. This is important today because many of our colleagues and students do not share the same cultural background as we ourselves. Accordingly, we cannot assume that we all hold the same ideas of what is 'right' and acceptable, let alone ethical. Sometimes we need to have the obvious laid out before us. I was once in a discussion at a Western university where a debate about signage in the toilets was underway. The locals (British) were outraged that anyone would need to be provided with a sign telling them not to stand on the toilet to use it — oblivious to the fact that in some non-Western countries, standing over the toilet (using its equivalent to ours) was the norm. Do not assume that everyone in the room understands and agrees with what you consider to be 'normal'.

Rights

The 'rights' of the individual that are recognised, valued and upheld in Western perspectives include concepts (or theories) such as autonomy, the opportunity to be able to give or withhold informed consent, voluntary participation, and protected privacy. There are more, but let's stick to these for now. However, in other societies around the world, the individual is not the unit of consideration but rather the family, the clan, the tribe or the village.

I have had great experiences in the past with African students who have had to abide by Western ethical codes, because they were registered at Western universities — but who had to implement local standards in order to succeed in their research at home. In many of their African villages, the Western demand to acquire consent to participate in research from each individual personally was typically over-ridden by the local custom of the village chief giving consent on behalf of the whole village. Everyone was therefore expected to participate whether they wanted to or not. Likewise, in many cultures a wife, daughter or sister cannot be interviewed or examined unless in the presence of her husband, father or brother. Try finding out about domestic violence in those circumstances! I once had an Arab student who assured me that there were no women with a drug problem in his country; this turned out to be due to the fact that his country only recognised and treated drug problems in men, so women apparently never suffered this way. A colleague of mine studying leprosy in India was astounded to hear a professional medical person tell her, at a leprosy clinic, that in India 'only men get leprosy'. The reality was that women with leprosy were forbidden to come to the clinic by their male relatives because all of the doctors were men, and a woman could not be examined by a male doctor at all, even with a male chaperone from the family. Therefore, apparently, 'women in India don't get leprosy'. Again, don't assume that we all see the world in the same way, even in the lab.

Given the prominence of Western influences on science, and on security, we need to recognise that so-called Western values are probably going to continue to dominate much of science discourse globally. Bearing this in mind, therefore, we need to continue to focus on the rights of the individual as a core element of ethical bio-chemical security.

The rights I have mentioned above for the individual (autonomy, etc.) may be seen as operationalising 'respect' for the individual — they provide a set of rules that together grant a level of respect for one person from another. As autonomy, privacy, and so on are also ethical concepts (developed as a means to protect individuals from various sorts of harm and to enable the maintenance of human dignity), they can form a useful part of an ethical framework. We can then use concepts such as these in constructing questions to assist us in formulating ethical pathways through our own work, as well as using them as a means to interrogate the ethical aspects of the work of others.

Responsibilities

The responsibilities of the researcher or the scientist towards the participant, or subject, or end-user of research are just as important. These include, in the West, the aim to do no harm; to provide benefit to the participant(s) where possible; to provide benefit to the wider public as a result of the research or work; to achieve and maintain a level of qualified professional proficiency in order to be able to carry out research or work properly and effectively; and to take responsibility as much as is reasonable for the outcomes of the research or work (this last item is highly debated and I seem to be the only person to date who adds this formally to the list of ethical responsibilities).

Who holds the ultimate responsibility for our work is therefore a key question. I raised the question in Chapter 1: 'If you win the Nobel Prize for the *good* outcomes of your work, is it right and fair therefore that you should be censured in some way for any *bad* outcomes of your work?' Bearing in mind some of the difficult questions that I asked in the Preface and in Chapter 1, where do you see the boundaries of your responsibilities for your work? In asking this, I am not only referring to the boundaries of your responsibilities once the work is in progress, but also the boundaries when the work is complete. In essence, I could ask you this:

- What do you *see* as your responsibilities during your work or research and after it has been completed?
- Can you identify the boundaries of your responsibilities in both time and space?
- Are you sure that you have got the right answers to this?
- Are you sure your boss and your employer agree with you?
- Are you sure that your colleagues, your staff, and your students agree with you?
- How do you know that the general public agree with you?
- Do not you think you had better check this out now?

Together, we can see that the rights of the individual and the responsibilities of the researcher or scientist (in all of his or her work, not just in 'research') form a powerful framework of duty and guidance. For the purposes of The Ethics Toolkit, this is the framework from which we will produce our 'applied' ethics in practice.

In rights and responsibilities, we now have the basis for a set of guidelines that will help us to decide what are the best things to actually *do* or not do, in order to 'be ethical' in our daily practice. As well as being a set of guidelines, our embryonic ethics framework can also be a useful road map to direct our security practice. By considering the concepts outlined as 'rights' and 'responsibilities' we can identify at least *some* actual or potential security problems; we may then use the same framework to show us some useful solutions to apply to these recognised problems. The framework can also serve as a tool with which to assess the work of others, helping us to easily identify ethical and security problems where before we would have not even noticed them.

The Toolkit framework is also a useful instrument for you to use to go back and review your own earlier work — remember what

Ian Ramshaw said about the Australian Mousepox experiment.[9] He recognised that he and his colleagues had been unwittingly engaged in potential dual use research without being aware of it. This recognition was a shock. Not only had they inadvertently created a transmissible virus that could overcome existing immunity conferred by vaccines, they had also created a virus that could result in sterility. How would you deal with a similar realisation?

I have met and worked with many scientists who have reviewed their own work and realised for the first time that they had previously been unwittingly engaged in dual use work — work that had the potential to be misused by others. Remember, I am not saying that we should not *do* work that has dual use potential, but that we need to be *on the lookout for dual use potential* and where possible, *take steps* to minimise the risks of misuse.

Descriptive ethics

You will sometimes hear people refer to descriptive ethics, which is also sometimes called 'comparative ethics'. This is generally what it sounds to be — a description of the ethical values of a specific group. This is relevant here because it is probably what you will engage in when you start observing your friends and colleagues at work — and yourself, of course. You will possibly be amazed when you see what you and your colleagues have been doing (or not doing) without realising the ethical implications (and not just in a science context).

When describing and comparing the values of two or more groups, we get 'comparative ethics' because we can compare and contrast differing values and views. Descriptive ethics focus on the following:

- the beliefs of people about values or 'what is right',
- What values are held in each cultural group — what is important in that group,

[9] Weir, L. and Selgelid, M.J. (2010) The Mousepox experience: An interview with Ronald Jackson and Ian Ramshaw on dual-use research. *EMBO Reports* 11(1), 18–24. Available on: http://www.ncbi.nlm.nih.gov/pmc/articles/PMC2816623/.

- How value-change occurs over time in groups,
- And so on…..

It does not aim to judge values, but simply aims to:

- *identify* them,
- *describe* them,
- and to some extent, *define* them (but remember that definitions may be wrong through misunderstandings).

Descriptive ethics is summed up in these questions, posed to an individual or a group or community: 'What do you think is *right*?' and 'What do you think is *wrong*?' In thinking about descriptive ethics you don't need to dig so deep that you get into asking 'what does "right" actually mean?', you just have to be able to say what this individual or group believe is 'right' and 'wrong'. You also need to be sure that you have understood them correctly and not let your own bias or misunderstanding affect your 'interpretation' of their values. The other issue here is in identifying what is *actually* believed rather than *what is said* to be believed. People are great at saying they believe in *x*, while their actions clearly show that they really believe in *y* (see my surgeon colleague above with the hand washing).

Normative (prescriptive) ethics

I'm including this here because normative ethics are concerned with asking 'How *should* people act?' *Normative* means the subject is *norms*. These are values that are considered 'the norm' by members of a given community, and all the members are expected to adhere to them (or have a good excuse not to do so). Normative values are typically those that have either been codified at an earlier time and handed down over generations, or have been adopted to manage certain situations. Usually, norms are so embedded in culture that they are not even recognised until someone breaches them. The Jewish 10 Commandments form a set of values that can be called norms (I know that is a religious example and I said I was not going to mention religion — but it's a clear example so I am using it).

Many cultures are governed by religiously-derived norms. Even secular societies tend to hold to norms that originated in religious traditions. The norm that says it is wrong to kill another person is widely held as a universal value and derives from religious values. Even so, we can see on a daily basis how this is disregarded. Consideration of norms often involves some element of judgement and criticism, along with considering the rightness or wrongness of actions from some stand point(s); what is a norm to you may not be to me. Other peoples' norms often look crazy, but we need to remember that we probably look crazy to the other people as well.

Often we hold conflicting values or norms without knowing it. We may state 'I believe that killing another person is always wrong.' Next day we may go to war with a gun and say that we are engaged in a 'just war.'[10] Hmm. Can you be right and wrong at the same time? There are two ways of looking at right and wrong in normative ethics, which directly relate to applied ethics (so we are going to use them in The Ethics Toolkit):

- *Teleological* ethics — argues that the morality (rightness or wrongness) of an action depends on the action's outcome or consequences,
- *Deontological* ethics — argues that decisions should be made considering duties and rights — and says that some things are *always* wrong or right regardless of the outcomes.

We'll come back to this below.

Confusion between norms and descriptions ('*Do as I say, not as I do*')

We cannot assume, assign or identify *a norm* from *a description*. This means we cannot identify a norm simply by saying what people *should* do from what we see people *actually* do. If I see you furtively changing something in your results, I should not infer that you think it is

---

[10] See Bagini, J. and Stangroom, J. (2006) *Do You Think What You Think You Think*. London: Granta.

acceptable to mess about with the recording of your results. On the other hand, if you know this is wrong but do it anyway, you are in effect demonstrating that you *do* have a norm — a norm of fiddling the results. You just like to keep your little norm hidden. Hmm.

As humans we are all fallible; we may be observed doing something that we know is wrong, but that does not stop us doing it. Just because it is being done does not mean it is "right" or that everyone agrees with that action/thought. Alternatively, someone could be doing 'wrong' and not realise it (see above, again, for my surgeon colleague with the taps) even if they *should* know better.

Relativism: are some more equal than others?

Ethics has something to say about this and it involves, awkwardly, a level of judgment. We cannot get away from this even if we want to. We live in a world today that focuses heavily on the notion of relativism. This concept is supposed to allow the 'equality' of all by viewing all value systems and traditions as 'equal'. This perspective is embedded in human rights, politics and international relations, and manifests in many Western societies as multiculturalism (amongst other indicators). Cultural relativism says that we should accept all cultural differences — which can mean behaviours, habits, attitudes, and beliefs — as being equally valid. This can be on a local, micro-scale, such as what you do in your lab, or on a macro-scale between communities or countries.

When it comes to science security, I have big problems with this, both as an anthropologist and a scientist. Relativism does not allow for questioning and a way forward in conditions of disagreement (feel free to disagree, but you should have some real examples, not just a vague feeling that I am wrong). Nor does it often accommodate the pragmatic aspects of life. If you are doing something that is harmful to yourself or others because it is always been done this way, do not you *want* to be told so that you can put it right? To accept all actions or beliefs and values as equally 'valid' is, in my opinion, not tenable if we want to have a get-along *secure* society with a common set of public values *in science* (I am not referring to wider society here).

We are talking in this context about values and practical actions that will protect the world from harm through the prevention of the development of biochemical weapons, so it seems to me that it is important to be somewhat definitive about what is right and what is wrong in terms of science security.

If you still prefer to hold on to relativism, think for a moment about this. Micro-organisms do not respect cultural values and behaviours. They will multiply and spread in any conditions that are 'right' for them. You cannot refuse to accept the existence of a pathogen or questions its validity, consequences or right to exist. This is where a lot of postmodernists will part company with me (bye then, see you at the hospital). You can declare your right to freedom from 'imposed cultural values that deny the totality of reality and the paradigms of the masses' as much as you like, but sooner or later you will reap the consequences. You may wish you did not have to abide by standard biosafety practices because you don't agree with them, believe in them or wish to go along with them; good luck in getting a job in that case. However, once in a position at the bench, if you become infected with anthrax, you have been infected with anthrax. You can not say 'But I do not accept the cultural or scientific validity of anthrax so it cannot infect me or kill me' (believe me, I have had postmodern social science students say exactly the equivalent to this). I have even had social science PhD students who refused to answer assessment questions on research ethics because they would not 'recognise the concept of ethics because it was the product of Western, Judeo–Christian narratives entrenched in privilege and bias.' Needless to say they were hard-core postmodernists. This is apparently some sort of privileged elite in itself. Had they been faced with religious fundamentalists claiming exemption from ethics on religious grounds, who knows what they would have answered. I sometimes wonder how postmodernists can get out of bed safely in the morning when, according to some of their claims, they cannot be sure that the floor is actually there. However, back to the real world. The fact is that I will respect your right to believe that the moon is made of green cheese, but I will not accept that you are correct, and you cannot expect me to agree with you or make me agree with you. Having said all this, which sounds very deontological, I also accept that ethics in practice is full of 'shades of grey', which is a teleological approach. Oh dear, back to the drawing board again.

While we need to *respect* the right of others to hold different opinions and values in their private lives, we still need to come to some sort of consensus about what is agreeable to the majority and do-able by all in the *public* sphere. In the case of good science practice, we need to agree on what is agreeable, or at least manageable, to *all* those who practise it, because science is practised in the public sphere (sometimes you need to compromise your personal values to some extent in order to get and keep a job). This means that the issue of bio-chemical security has to be taken up by leaders and experts in science, rather than non-experts. In effect, I am saying, and professional associations already assume this, that if you do not agree to 'do practice' this way, then you cannot be recognised as part of the profession. You may not agree with the principles, but in the context of work you have to *live with* the principles, or you cannot work in the sector. Codes of Practice and Codes of Conduct are built on such foundations. I firmly believe that if we do not grasp the opportunity to manage bio-chemical security ourselves as scientists soon, someone else will do it for us. This should concentrate our minds.

In order to participate in science, we have to agree to certain norms (biosafety practice norms, for example) being the basis for our practice. This may require us to compromise on some of our private values, but in order to work together, we agree to work to a common set of public values — or 'ethics'. Do not get your private values mixed up with your public ethics. They may both be the same, or they may be different. If the latter is the case, internal conflicts will arise and you will need to do something to reduce the effects of such conflict, for your own wellbeing.[11]

---

[11] Sture, J. (2013) Moral Development and Ethical Decision-Making, in B. Rappert and M. Selgelid (eds) *On the Dual Uses of Science and Ethics: Principles, Practices, and Prospects*, Canberra: ANU E-Press, pp. 97–120. Available on: http://press.anu.edu.au/?s=On+the+Dual+Uses+of+Science+and+Ethics. Accessed 24/11/15.

Academic/Scientific Freedom and Ethics

Bio-chemical security is an ethical issue. You might not agree with this. You might think that it is all a fuss about nothing. Perhaps you remain convinced that your work is not a bio-chemical security risk now and won't be in the future. However, times are changing. It is at this point that the 'biosecurity police' come up frequently against the 'academic/scientific freedom' argument. This is not the place to debate academic/scientific freedom in detail, but I do want to address it briefly here.

Yes, you should, in theory, be free to go where the science takes you. You should be able to publish everything. Yes, you do have skills and knowledge that can benefit humankind. Building on the shoulders of giants has never been easier. But, in today's world of instant communication, easy global travel, extremely effective and efficient scientific practices, equipment and facilities, many of which have the ability to fundamentally change the very materials of life, you also need to consider all of these benefits from the perspective of those who would seek to misuse them.

Personally I do not believe that beneficial science needs to be 'stopped' and 'dangerous' work prevented. What I do believe is that in order to go ahead with such work, extra steps need to be taken to protect such work — and to be *seen* to protect such work — from those who may misuse it. This is just a common sense approach that protects scientists from censure as much as it protects the public from danger. Unfortunately, arguments for restrictions on scientific work and publication for security reasons are now the cause of much antagonism between scientists and the security community. I have heard and read comments from many scientists who are outraged at any suggestion that the (relatively) unfettered practice and full publishing of their work should be restricted in any way due to security concerns. In the Preface, I mentioned Einstein's quote: "The right to search for truth implies also a duty; one must not conceal any part of what one has recognized to be true." This has been the basis for scientific reporting since the 17th century Enlightenment — in other words, well before Einstein said this. Quotes such as this tend to be aired a lot in the academic/scientific freedom argument. But can we still stand on this foundation so easily today? I believe that it is a good foundation, and a worthy aim, but in today's real world we may need to be prepared to look at the occasional compromise of our professional values and traditional practices.

Academic and scientific freedom is often seen simply as the right of all academics and scientists to do whatever work they want to just because they *can*, plus the right to proclaim their achievements widely through publication. It is also, under this argument, the right of the public to have access to all scientific knowledge and advancement. That's all good and in an ideal world I would be at the front of the demonstration waving a banner to support this. But we do not live in an ideal world.

I used to have more respect for the 'freedom' and 'rights' approach before I saw more than a few of the outraged scientists apparently wanting to selectively restrict this freedom and these rights whenever they wanted to apply for a patent or prevent a competitor from 'stealing' their idea until they are ready to publish it (for which read, 'ready to make money from it'). Strangely, when commercial gain looms across the horizon of academic and scientific freedom, the rights suddenly change somewhat. When promoting ethics approval processes and ethics policies on the publication of postgraduate thesis and dissertations in a range of universities and research institutions, I have often come across the anxiety of scientists, PhD supervisors and science educators who see open and full publication as a threat to their economic advancement — two, three and five year embargoes on publication have all been proposed and implemented for work that could make some money for those involved. Hmm. Where is academic and scientific freedom and the right of the public and other scientists to know about it all in those circumstances?

Whatever you think of the freedom and rights argument, what will happen when your scientific work ends up as the 'cause' — the foundation of a biological or a chemical weapon? Think back to what we discussed earlier in Chapters 1 and 3 about the potential adverse outcomes that may be suffered by you, your institution and your employers if your work is misused to produce some sort of biochemical weapon (of any sort). Think again of the Italian seismologists who were given jail sentences for not carrying out their perceived duties 'correctly' in 2012. The verdict that resulted in them receiving jail sentences was based on how they assessed and communicated risk. Even though this verdict was later overturned, this is arguably not an issue that can be ignored by scientists of any discipline today.

Bio-chemical Security ethics

I do not have answers to all of the questions raised by the security issues that now face science. But I do believe that by implementing an applied ethics of bio-chemical security in the practice of science, we will be able to better identify and mitigate the risks of misuse of our work. Science is not our professional possession. Most of it is publically-funded, so technically science 'belongs' to all of us. We are just the bearers of it for the next generation. We should also be guardians of it for the next generation.

Deontology and Teleology

Let us look here at the two approaches to ethics that I mentioned above under 'Normative ethics'. You will come across these two concepts if you decide to read up on the philosophy of ethics (which is actually interesting if you have the time to do it). It is worth spending a few minutes getting these clarified in your mind as I have used them after each section in the next chapters of this book. It is useful to think of them being at opposing ends of a continuum — which is overly simplistic but a good place to start.

These words are derived from the Greek:

Deon — 'duty', 'obligation';

Telos — from the root meaning 'end' or 'purpose'.

I am going to simplify here, so apologies to any readers who are philosophical ethics experts. This book is concerned with an everyday practical version of ethics, so I am not going to dig into detailed debate or the philosophical technicalities of ethics.

Very broadly, deontological ethics holds that people must follow 'the rules' no matter what the outcome. For a deontologist, a situation can be assessed as either good or bad depending on the rightness or wrongness of what was *done* that brought the situation about. Deontology is about being bound to rules and doing your duty according to them.

In this approach, certain actions are *always* wrong or *always* right, depending on what the rules require. Examples might include 'you must always tell the truth no matter what the outcome' or 'thou shalt not kill' (religion again!). It is probably easiest if we think of this perspective in the bio-chemical security context as saying 'do not ever do this' or 'always do this,' depending on what we are looking at. So we might say 'you must always abide by biosafety regulations' or 'you must never switch the taps off with your hands in the operating theatre after scrubbing up.' (see above). These are ethically *deontological* acts. Think of deontological perspectives as black and white. There are no shades of grey here.

This approach can, of course, lead to bad outcomes when the doing of a "bad" act might have saved some people from danger or harm. However, the general idea may be simplified by saying that certain actions are always wrong or right, no matter what good or bad outcome may emerge from them being carried out. In this case, you could allow someone to die slowly and painfully if you believe that giving them that last shot of diamorphine would 'see them off', and you do not believe in killing. The contradictions in this are huge, but it is just a simplistic example. To you as a dyed-in-the-wool deontologist, the patient dying in unnecessary pain could be said to be a 'good' situation because you are not doing a forbidden act (killing him, even indirectly). The fact that the patient is in pain is to the deontologist arguably irrelevant. This is one of the core arguments in the 'mercy killing' debate or the 'assisted suicide' debate. It is also used in the anti-abortion debate. Many conscientious objectors in the two world wars of the 20th century held to this belief — I will not kill. They paid for it socially and economically as a result. Of course, any reasonable deontologist would also see the arguments in favour of humanity, sensitivity and kindness and that leads us into the teleological minefield where we have to look at consequences just as much as at actions themselves.

Teleological ethics holds that an action, even certain "bad" actions, may be justified *if the outcomes are good* for some or most people. Another way of saying this could be 'the end justifies the means'. You will find that this approach is often referred to as *consequentialism*, as it is concerned with consequences or outcomes of actions, rather than the value of the actions themselves. *Utilitarianism* is an example of a teleological perspective: the best outcome is the one that provides the greatest good for the greatest number of people. However, this does not protect the fewer people for whom the outcome is not good. One of the problems with a teleological approach is that, arguably, insufficient account is taken of the suffering of those for whom the outcomes are not good. In practice, however, the teleological approach is typically taken in public life — politicians do this all the time, as a means of appealing to the greatest number of people (for 'people' read 'voters'). Teleology has many shades of grey.

Before you get too fond of the teleological perspective, bear in mind that in bio-chemical security we are going to need some deontological input as well. It is all very well saying that we have to be reasonable and sensitive, but we also occasionally need to put hard and fast rules in place for the greater good. That is deontology. The 'grey areas' come under teleology, and the weakness in that is the tendency to bring subjectivity and personal opinion in to replace objective ethical debate.

These definitions are highly over-simplified here, but I am providing them in this elementary format just to introduce you to them. In order to provide a simple way of seeing this, I have presented the two approaches as opposite ends of a continuum in Figure 4.1. This again is an over-simplification, but it is a start and simply meant as an illustration to help you to start thinking about these concepts.

If you spend a few minutes thinking about your answers to some basic ethical questions, you may find that your views are not as

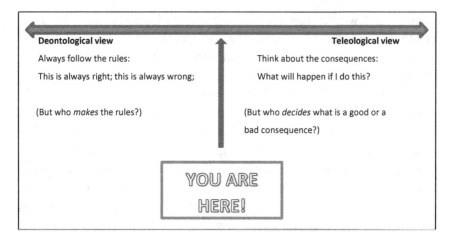

Figure 4.1:   The Ethical Continuum: which way will you go?

straightforward as you previously thought. For example, you may be against killing other people, but how would you answer this:

'Can I legitimately murder a person who is going to murder my children if I do not kill him first?'

Some questions you could ask yourself when reading this book include:

- What might a bad action or outcome be in terms of my research or work?
- What might a good action or outcome be in terms of my research or work?
- Who defines it as good or bad?
- Who is affected by the outcomes of the action:
  - Now?
  - Later? (some unspecified time in the future)
  - Here ? (where you are/live/work)
  - There ? (somewhere else)
- Who or what might influence whether or not an action or outcome is facilitated?
- Where is the power and authority in this decision process?
- What would happen if this power relationship changes?

When you read the various sections on ethical principles in the following chapters, keep coming back to this diagram and thinking about where on the continuum you might place your own decisions and opinions. Where you locate your answers on the continuum will probably vary according to the importance of decision (in your own mind), your ability to affect the outcomes, your background and your personal moral value set as well as *who is affected by the action*. What is obviously right and wrong to you may not be obviously right and wrong to others. Why? What pressures are acting on you and them that result in differing opinions and values?

You may find that your *opinion* sits in one place but your actions sit in another. This means that you have got some sort of tension going on in your mind — you may need to do something to fix that or it will only get worse.[12] Letting ethical tensions brew is not a healthy situation for you, your colleagues or your employers. If you find that your work is challenging your personal values, even if your work is 'ethically approved', you will not be able to work to the best of your ability and your work will suffer — as will your career if you are not careful. I have used this example elsewhere, but it bears repeating: if you are Catholic you won't want to be in charge of an abortion clinic (religion again!). Of course, you may be a Catholic who picks and chooses and disagrees with the teachings of Rome, in which case you *may* choose to hold such a position — albeit at a social cost in your religious social circles. None of this is straight-forward until you get used to the arguments. Likewise, if *your* job requires you to engage in work that you fundamentally disagree with, or have major concerns about, the situation will only end badly, either for you, your employers or all of you.

---

[12] For an example of ethical tensions in a scientist, take a look at my chapter: Sture, J. (2013) Moral development and ethical decision-making, in B. Rappert and M. Selgelid (eds) *On the Dual Uses of Science and Ethics: Principles, Practices, and Prospects*, Canberra: ANU E-Press, pp. 97–120. Available on: http://press.anu.edu. au/?s=On+the+Dual+Uses+of+Science+and+Ethics. Accessed 24/11/15. This chapter contains a lengthy example of an 'ethically conflicted' scientist.

Are there Any Ethical Issues in My Scientific Work?

Answer the following questions honestly please:

- Do you work with people?
- Do the outcomes of your research *affect* other people?
- Does your work and its outcomes *involve* other people?

If you answer 'yes' to one or more of these questions, you have ethical issues in your research.

Here is a quick exercise for you. Get a pen and paper, and jot down your answers to the following questions.

1. Think of an event in your lab that involves or causes a biosafety problem — it can be anything at all, as long a biosafety problem arises from it. Jot down a short explanation of the problem as follows:
   a. What is the nature of the problem? (e.g. a spill, a failure of equipment etc.)
   b. What or who caused it?
   c. How was the situation recognised as a biosafety problem?
   d. What can be done to fix the consequences of the problem?
   e. What can be done to prevent it happening again?

That should not take you too long. Next, jot down your answers to these questions:

2. Think of an event in your lab that involves or causes an ethical problem — it can be anything at all, as long an ethical problem arises from it. Jot down a short explanation of the problem as follows:
   a. What is the nature of the problem?
   b. What or who caused it?
   c. How was the situation recognised as an ethical problem?
   d. What can be done to fix the consequences of the problem?
   e. What can be done to prevent it happening again?

Why was it easier to answer the first question than the second? Because you operate in a framework for biosafety that provides you with concepts (for example, containment) and actions (such as standard operating procedures, testing of equipment regularly, wearing PPE, and so on) to implement the framework. Plus, you are *familiar* with biosafety theory and practice, so compliance is now automatic for you. You *understand* and *recognise* biosafety problems. If you had an ethics framework with which you were equally familiar, you could answer Question 2 as easily as Question 1. Well, you are reading this book, so you're getting there.

Ethics as Responsibility for Yourself as Well as for Others

Here is something else to consider as well. While I refer to 'other people' here, do not forget that you also owe an ethical duty of care to yourself. A number of scientists in the past are recorded as having tried out experiments on themselves (even Nobel laureates) but in today's litigious society, this is probably not to be recommended. However, if you are determined to be 'unethical' towards yourself in the name of science, go ahead, on the understanding that you may well lose your job — or win a Nobel Prize. Let us look for a minute at the case of Werner Forssmann, who carried out some self-experimentation in the summer of 1929.

Forssmann was a German doctor who pioneered the technique of cardiac catheterisation. This technique is now widely used as means of assessing the condition of the heart for diagnostic purposes or alternatively, for therapeutic purposes. It involves the passing of a thin catheter into the chambers of the heart, guided by fluoroscopy and entering the circulatory system by a vein, typically in the groin.

Forssmann had seen a print in a 19th century medical text showing the cardiac catheterisation of a horse, in order to measure the changing pressure of the blood in its heart. This fascinated him and he determined to try this himself, using the cubital vein (in the arm). He discussed this with his boss, who told him he could not self-experiment but should do animal experiments first.

Forssmann then went to the nurse in charge of the operating theatre, who controlled access to the necessary surgical instruments. Two weeks later she agreed to support Forssmann in his procedure and apparently volunteered to be the subject of the experiment. On the day of the procedure, he persuaded the nurse to lie on the operating table to 'counter any side effects of the local anaesthesia' in her arm. When she lay on the table, he strapped her arms and legs in place to prevent her moving. He managed to prepare his own arm with local anaesthetic without the nurse noticing and prepared her arm in the same way, as if he were about to do the procedure on her. However, he then inserted the catheter into his own arm. He then asked the nurse to call the x-ray staff. It was only then that the nurse noticed that he had inserted the catheter into his own arm and not hers (her arm was numb from the local anaesthetic and she had thought he was working on her arm).

He released her from the table and together they went downstairs to the x-ray unit. An x-ray nurse put Forssmann behind the fluoroscope before one of Forssmann's colleagues rushed in and tried to take the catheter out of his Forssmann's arm. Forssmann successfully pushed the catheter up into his own heart and had an x-ray film taken of it. He was in due course summoned before his boss, who he managed to convince of the usefulness of the exercise by showing the x-ray film. However, worryingly by today's standards, he was then given permission to try the procedure again on a terminally ill female patient in the hospital. He apparently administered medication to the woman via a catheter directly into the right ventricle, having put it into place without fluoroscopy but by visually estimating the position on the woman's body surface (!).

In November 1929, Forssmann published a paper on his experiences in 'probing the heart' but due to the media attention, was forced to leave his job for a while. He later went on to join the Nazi party and became a Major in the German army until he was captured and held in a POW camp during the Second World War. While he was imprisoned, his paper was read by André Frédéric Cournand and Dickinson W. Richards, who went on to hone his technique. They later shared with him the Nobel Prize. When he heard the news of his award, Forssmann said: "I feel like a village parson who has just learned that he has been made bishop."[13] (Religion again!).

I barely know where to start with the ethics of this example, despite the ultimate medical value of the findings. Firstly, we have the issue of Dr. Forssmann in effect coercing, or at least pressuring, a professional nurse to assist him in an unauthorised procedure; not only was it unauthorised, it had been previously actively forbidden by his boss. This is ethically problematic not only in terms of Forssmann's own position in the hospital and in his profession, but also in terms of the loss of, or pressure on, the nurse's personal and professional autonomy. According to Heiss,[14] the nurse took two weeks to 'agree' to assist Forssmann in his experiment. In the 'giving' of her consent under some pressure (so maybe no true consent at all) and in view of the fact that in putting her in this position, Forssmann was exposing her to various sorts of harm including possible death (in her own mind), we must question the whole assertion of her voluntary participation. She *may* have been fully informed and given full and free consent, but given the hierarchies in hospitals in the early 20th century, I do not believe this can be certain.

We may surmise that Forssmann did not in fact fully inform her of the risks of the procedure — even in a hospital power relationship in the early 20th century it seems unlikely to me that any nurse would have 'consented' had she known the real risks. If she did consent even knowing the full risks, we can question this in itself — she may have felt pressured into agreement. Whatever the true situation, as a power relationship was involved here, autonomy and consent were compromised.

Secondly, the nurse, under whatever consensual circumstances, acquired or made available for use the necessary equipment for an

---

[13] Collins, N. (2013) Nobel Prizes: Winners who experimented on themselves. *The Telegraph*. Available on: http://www.telegraph.co.uk/news/science/science-news/10360202/Nobel-Prizes-Winners-who-experimented-on-themselves.html. Accessed on 24/11/15.

[14] Heiss, H.W. (1992) Werner Forssmann: A German Problem with the Nobel Prize. *Clinical Cardiology* 15(7), 547–549. Available on: http://onlinelibrary.wiley.com/doi/10.1002/clc.4960150715/pdf. Accessed 28/12/15.

unauthorised purpose, for which permission had already been refused by the hospital authorities — the hospital that employed her and on whom she depended for her income and future promotion. So she was ethically compromised even in just enabling Forssmann to do the procedure.

Thirdly, we find that she amazingly 'offered' to be the guinea pig — again, presumably due to the power relationship inherent in the situation. This contravenes every ethical concern in the book — because of the power relationship (which of course looks far worse now in the early 21st century). The ethical problems in being the guinea pig include the loss of her autonomy, a lack of voluntary participation (the power relationship negates her 'consent' to a great degree), the exposure to harm (physical, such as possible death; economic if she loses her job), loss of physical privacy and autonomy (her body would have been 'invaded' as part of the procedure had Forssmann actually experimented on her), a lack of beneficence to her (what could she have gained by all this?) and so on.

Fourthly, he strapped her to the table even though she had asked to sit in a chair. This was further deception on his side, as he was strapping her to the table so that she could not prevent him inserting the catheter into his own arm (which he must have surmised she may do). He then broke his word, through which we can assume he gained her cooperation, and proceeded to experiment on himself. He anaesthetised his own arm unbeknown to the nurse, while also preparing her arm with a local anaesthetic. He then inserted the catheter into his own arm. While avoiding the risks of experimenting on her, Forssmann now put her further in harm's way by leaving her in the stressful situation of being complicit in a senior doctor carrying out a procedure next to her which could kill him. If he died, she could not help and when would she be found? What about her future in the hospital then?

Fifthly, he released the nurse from the table and they both went to the x-ray suite downstairs. Now he was asking the nurse to help him downstairs with a catheter in his arm, with no assurance that this would be safe for him. Once there, a colleague of Forssmann's tried to remove the catheter from Forssmann's arm, but Forssmann 'overcame him,'[15] indicating some sort of struggle. Using fluoroscopy an x-ray film was produced showing the catheter in Forssmann's heart. The staff in the x-ray suite were also thus compromised.

There are even more 'in detail' ethical issues here, including the further experimentation on the female patient, but these are surely enough to start with.

Let us look at the outcomes for the hospital and for Forssmann. The paper that describes these events states that the hospital was concerned with the attention being paid to it by the media, implying that this was of more concern than the unethical behaviour of Forssmann. It also describes how Forssmann was required to leave the hospital because he had not asked his new chief for permission to carry out the experiment (power relationships again?).[16] Forssmann did in fact return to work and went on to have a successful medical career. I will leave aside the fact that he later became a Nazi and look ahead to when his work was recognised by others. He eventually won the Nobel Prize in 1956 with others who went on to develop his work.

What do you make of this? While this is a medical example, and one step removed from the laboratory bench, there are still some very useful examples of unethical scientific behaviour on the part of several professionals. Do you have any power relationships in your lab? (I'm sure you said 'no'). Are you sure you don't? How does this make you feel about your ethical duty of care to yourself? The key point here is that it is actually quite challenging to be unethical towards yourself without being unethical towards others at the same time.

Even had Forssmann managed to do the experiment on himself without involving the nurse or the x-ray staff, he would still have had to get access to the equipment. This has implications for the security of the equipment. You may say that the operating theatre staff and the x-ray staff could not have been expected to know that someone would gain access to facilities and equipment for unauthorised purposes. This is arguably not a safe assumption to make today. Hence the need for an ethical framework that we can use to identify and respond to

---

[15] *Ibid*, p. 548.
[16] Heiss, H.W. (1992) *op. cit.*

risk, while at the same time protecting ourselves from accusations of insufficient attention to bio-chemical security.

This example will probably be seen by many readers to come under a teleological approach to science. The end justifies the means. The development of cardiac catheterisation followed on from this experiment, so the experiment was worth it in order to achieve this good outcome. Many scientists think that this is acceptable. Many others disagree and would place this example under the deontological approach — don't ever do this. You can see more of such a situation in Chapter 8 with the Thomas Butler case. None of this is easy, but that's life. We have to make the best of a bad situation. Is that deontology or teleology?

Chapter 5

The Ethics Toolkit 1: Ethics as Other People's Rights

So Where Do I Start?

Until you are familiar with the ethical principles, and with applying them, the easiest way to consider them and to apply them is by asking questions about each principle. The questions relate to the work you or those under your oversight are about to carry out, or are already working on. You may also find it useful to apply The Toolkit retrospectively to see where things went wrong, or may have gone wrong, in the past. Each question below relates to your relationship with other people in terms of *their* rights. The next chapter deals with *your* responsibilities as a researcher or other science professional. You will see that there is some overlap between the rights and the responsibilities as you read on.

I have divided the questions up into sections, each focusing on one ethical principle. Once you get used to asking such questions, and of course, with your familiarity with your own work and that of your colleagues, you will be able to come up with even more relevant questions in your own context. Think of all these questions in terms of *people, materials* and *information.* In bio-chemical security, we also need to think about protecting things, knowledge and information from 'harm' — actual misuse. I talk below about *people,* but you can apply the same ideas to the science as well.

Who Do I Need to Protect?

The first simple key is to consider *who* may need protection from some aspect of your work, should it be misappropriated for hostile dual use purposes (you can also consider plants and animals as well as people). To put it bluntly, *everyone* needs protection from the misuse of your work in terms of its possible misuse in the development of biochemical weapons. I know that this is an unthinkable concept to most readers, but in today's world we need to start thinking about the unthinkable. Having said that, trying to think in the large-scale, intangible 'unthinkable' way about your work is not particularly helpful when you are looking at a blank page. Let's start by thinking of this in a more manageable way.

Think back to the 'ripples in the pool' diagram that we saw in Chapter 1. I have reproduced it here to refresh your memory. This is a good place to start because it condenses 'everyone' into readily clear, recognisable, and initially, 'manageable' groups. But at what point do the effects (the ripples) become unmanageable?

If you start with the 'ripples' closest to you, you will find that the first affected people include you as the scientist (or as a person working with scientists), your colleagues, your family plus those of your colleagues, plus all their work associates; any research participants or donors of data-generating material; *plus* those associated with them.

YOU ARE HERE

> You and your actions
>
> Your internal colleagues and their actions
>
> Your external work contacts and their actions
>
> Your family and friends and their actions
>
> Colleagues' families/friends and their actions
>
> Local population and their actions
>
> Wider population (up to national)
>
> Global population (international)

Figure 5.1: The Ripple Effect Diagram — where would you place the actions of each group of people after yourself, and where would the effects (the ripples) end?

This goes way beyond what you would normally consider in terms of *biosafety* protection. As well as these, we need to consider visitors to your facility; people involved in delivering to your facility; those removing waste and end-products from your facility; local residents and workers and finally, the rest of the general public right up to the national and global scale.

When we take the dual use awareness approach — the bio-chemical security approach — we take the wider view and consider the general public, locally, regionally, and nationally (or beyond) *as a first concern*. This is the opposite of biosafety, where the first concern is with the safety of the individuals coming into contact with the potentially dangerous materials — in other words, the protection of people in the lab. With bio-chemical security, the first concern is the safety of the general public, which involves the safety of the people in the lab and *also* the protection of the work in the lab from opportunities for misuse. You have to think beyond the laboratory door here.

This means that you may have to start thinking about your work in a slightly different way. The bio-chemical security sort of containment is wider than biosafety containment. A useful way to think of this is as a *social responsibility* approach to your work. I am *not* saying that there is no social responsibility inherent in your practices already — clearly there is, or you would not be engaging with biosafety and all the usual regulations that protect you and your colleagues. However, bio-chemical security involves a wider sense of social responsibility — the protection of whole populations, not just you and your colleagues in the lab. This may seem either overwhelming or unnecessary at first. Overwhelming, in the sense that you cannot be personally and professionally responsible for the protection of everyone from all possible misuses of your work. Unnecessary, in that you probably feel that you are already overburdened with oversight and regulation. Plus, you are probably still 99% sure that your work is not a danger to anyone. This is where we need to come back to what we looked at earlier; you cannot *eliminate* risk, but you can *reduce it* and take steps to *mitigate* the risks and any adverse outcomes. This is where ethics comes in. Asking ethics questions is a way to reduce and mitigate bio-chemical security risks.

What am I Protecting Them From?

The second simple key is to ask *what* may these people (or plants and animals) need protection *from*? What uses or outcomes may arise from your research or work, or be facilitated by your research or work, that could be used for hostile purposes by individuals or a group? Remember that dual use — misuse — can be facilitated unintentionally as well as purposely. This is the level at which dual use risk considerations need to be assessed as a means of achieving bio-chemical security. Being aware of the potential for dual use — hostile misuse — of your work is simply taking a common-sense approach to protecting people from consequences that you do not intend to follow on from your research or work activities. Common sense tells us that we cannot be totally responsible for *everything* that may ever emerge from our work across time and space, but equally, ethical responsibility requires that we take *reasonable precautions* to minimise the risks of *foreseeable* problems as much as we can *at the time of undertaking the research*. Remember that we are talking here about the wider world beyond the laboratory door and beyond the intended 'target audience' for the beneficial aspects of your work (such as patients with a specific condition who will be treated by your new drug and so on).

Mitigation in Advance

A third point that can motivate you to look at the potential for misuse that could arise from your work is the issue of prepared responses once misuse has occurred. If you look at the Lemon–Relman Report[1] you will find their fifth conclusion states that no matter what steps are taken to prevent the hostile misuse of science, you need to recognise the 'virtual inevitability' of the misuse of new technologies in the life

---

[1] The 'Lemon–Relman Report' — National Research Council Committee On Advances In Technology And The Prevention Of Their Application To Next Generation Biowarfare Threats (2006) *Globalization, Biosecurity, and the Future of the Life Sciences* Washington DC: National Academies Press, p. 256. Available on: http://www.nap.edu/catalog/11567/globalization-biosecurity-and-the-future-of-the-life-sciences. Accessed 30/12/15.

sciences and that you need to plan for 'rapid and effective' responses when these events occur.

You will be therefore welcomed when you have not only identified some potential misuse of your work but have also devised a way(s) to mitigate the adverse outcomes if and when misuse occurs. In other words, you are devising a response to a future problem. Hopefully your response will not have to be implemented, but it is there if needed. An example of a mitigating action would be the development of a vaccine if you are planning to devise a 'new' pathogen using synthetic biology, or altering the transmission route of a pathogen.

Of course, when you are seen to be taking these kinds of precautions, you are likely to prompt other colleagues to start to think along these lines with their own work. In this way, good practice will spread. Once the benefits of this approach are recognised, more people will want to join in. As we have noted in an earlier chapter, it will only take one big security incident to concentrate our minds on how to avoid repeats. By taking an ethics approach to bio-chemical security in pursuit of a socially responsible science, we can at least minimise and mitigate the risks of our work.

All of this highlights the necessity for scientists to be educated in ethics from high school up, as an ongoing activity. It is crucial for all scientists to be challenged with ethical debate pitched *at the level at which they are currently working.*[2] Research has shown[3] that ethics tuition and learning is most effective when learners are confronted with ethical issues that reflect the kinds of ethical challenges relevant to their current levels of responsibility. In addition, the same research shows that a loss of early high standards of ethical behaviour tend to occur if juniors see seniors (in any context) acting in unethical ways. This leads to 'ethical erosion', and the loss or dilution of ethical standards, which tends to spread 'down the chain'. It is therefore of the

---

[2] Sture, J. (2010) Educating Scientists about Biosecurity: Lessons from Medicine and Business, Chapter 2 in B. Rappert (ed) *Education and Ethics in the Life Sciences: Strengthening the Prohibition of Biological Weapons.* Canberra: ANU E-Press. Available on: http://press.anu.edu.au/titles/centre-for-applied-philosophy-and-public-ethics-cappe/education-and-ethics-in-the-life-sciences/. Accessed 28/12/15.
[3] *Ibid.*

utmost importance for seniors to demonstrate high ethical awareness and standards at all times, in order to maintain and promote the same approaches in those following their lead.

The follow-on from this is that a 'suitably qualified' scientist should not only be suitably qualified in his field with the appropriate degrees and experience, but should also be suitably qualified *in ethics*. A scientist with a full grasp of ethics in relation to his or her work is a scientist who can defend him or herself *and* the work. Ethics is defence. It is as simple as that. If you have planned and carried out your work with ethics as a first consideration and as an ongoing yardstick underpinning your decisions, then challengers will have to challenge the ethics, not just you. In cases of disagreement, it is then that the ethics that can be reviewed, tightened up or loosened off, rather than you necessarily being sacked or disciplined.

Let us look now at the actual ethical principles we need to address in relation to a social responsibility approach to research. The first group relates to the rights of others; some responsibilities fall on you in order to protect the rights of others. We will look at your *direct* responsibilities as a scientist in the next chapter.

The Autonomy Principle

This principle is based on the right of every human being to self-determination. Of course, not all cultures accept this as valid, but as we noted earlier, 'Western' values tend to dominate science and look to do so for the foreseeable future, so let's run with this. The right to self-determination means that individuals should reasonably be able to make decisions affecting themselves and their wellbeing without being subject to deception, coercion or other forms of pressure. It also means that people increasingly expect to be made aware of issues concerning themselves and their wellbeing so that they can make appropriate choices for themselves.

The modern world has seen a significant rise in the rights of the individual when compared to previous eras. Society, on a global scale, has seen many changes in attitudes to authority, power and moral values even in the last decades, let alone centuries. Individuals, groups and communities now routinely take legal action to protect themselves or to seek redress for perceived or actual wrongs done to them. All of these issues indicate the value that society places on the autonomy of the individual, the group, community, and population. In the research and science context, this autonomy is preserved, as far as possible, by the implementation of three principles: voluntary participation, consent, and respect for privacy (such as confidentiality). While these are very socially-driven values, you already accommodate them in science, even if you do not realise it. I am suggesting here that these values are going to become even more recognised in the practice of science, and will therefore require greater responses from scientists.

If you want to think about some recent examples of scientists being called to account around these issues, just consider the Alder Hey scandal in the UK,[4] the H5N1 debate[5] in the scientific and mainstream media and the professional challenges thrown at the Italian seismologists after the Aquila earthquake.[6]

Voluntary participation

You may well be wondering what this has to do with scientists and laboratories. After all, do not people who work in and for laboratories *choose* to do so? Yes, of course they do. But do they all fully understand and agree with what is being done in those laboratories? Do they need to? I would say that yes, they do need to know. Support staff may not need to know the technical details of the work, but in today's climate it is highly advisable for all staff to know what is going

---

[4] See Hall, D (2001) Reflecting on Redfern: What can we learn from the Alder Hey story? *Archives of Disease in Childhood* 84, 455–456. Available on http://adc.bmj. com/content/84/6/455.full. Accessed 8/12/15.

[5] See *Nature* special edition 'Mutant Flu'. Available on: http://www.nature.com/ news/specials/mutantflu/index.html.

[6] See Nosengo, N. (2012) Italian court finds seismologists guilty of manslaughter *Nature* 491, 15–16. Available on: http://www.nature.com/news/italian-court-finds-seismologists-guilty-of-manslaughter-1.11640. Accessed 27/12/15.

on in the lab for which they work. It is only good practice for all staff to be able to choose whether to join or continue to work in a lab that is working on something actually or potentially controversial.

Beyond the staff and support workers, we also need to consider the 'donors' of the materials used. If, for example, you are working on human tissue, as well as accommodating the demands of all the relevant national legislation, you should also consider what implications your work may have on the individuals or groups who have donated tissue or other materials for research. Just because 'Manonthestreet' has donated his blood, or has agreed to let you keep his bits and pieces after surgical removal, is it right to simply go ahead and 'do what you want' with these materials? What if a 'donor' does not wish his tissues to be used for certain sorts of research? We need to think a lot more about this sort of thing today, not just in hospital settings but in working labs and educational labs too. Since the Alder Hey incident in the UK, British consent issues have been tightened up significantly to avoid such 'reuse' scenarios arising without patient/donor consent. This can lead to perceived delays or gaps in research that could have gone ahead earlier with fewer restrictions. Nevertheless, this is necessary to respect consent.

If your lab has a regular source for materials such as human tissue, are you satisfied that the collection and consent procedures used by that source are good enough? Even legally-controlled oversight and procedures can fail due to human error. If something goes wrong in the future, can you excuse yourself from any responsibility by referring complainants to the source of the collection and consent, rather than standing by your own assumptions that 'we do not have to do this bit'?

While most people working with scientists in commercial organisations, universities and other research facilities have a pressing economic need to keep their jobs, in today's climate of human rights and the primacy that society places on 'choice', I would argue that, where possible, all facility workers, scientists or not, should be fully informed, or at least reasonably informed, of what sort of work is being carried out in any facility. In this way, they can decide for themselves whether they want to be associated with the work and by implication, exposed to the possible consequences. In other words, we need to recognise that *everyone* in a facility is involved in the scientific work and outputs of that facility, whether they are the chief, the secretary or the cleaner. They all need to voluntarily participate in the science going on in the lab.

Further, by doing so, not only are employers protecting the rights of their employees, but they also are more likely to recognise those who fundamentally disagree with what is being done, or who may be the source of problems later. As I said earlier, if an individual is engaged with work that challenges his or her personal values, something is going to break sooner or later. It is therefore simply a matter of good people-management to be open about the nature of the work being done in a particular facility.

Voluntary participation is particularly relevant in the recruitment and retention of research participants; it also has considerable overlap with consent issues. Obviously this also applies to the recruitment of scientific staff, trainees, and support staff. It also applies to groups or populations of people in terms of what they are exposed to (participation in research or the outcomes of it can be involuntary and this needs to be a major consideration in the dual use context). Consider, for example, how a community may feel if its members have no control over fluoride being added to their water, or vitamins added to their bread. What else may be imposed on people with or without their knowledge, over which they have no or little control?

Why 'voluntary' participation?

As a general rule, all research participants should take part in research as volunteers of their own free will because it is generally accepted by society that adults have the right to be self-determining individuals.[7] This same consideration needs to be granted to staff in science facilities. As such, they should be free to decide what they engage with. At the population level, in the context of the possible dual use of biotechnology in

---

[7] Those in the military do, of course, voluntarily give up a degree of autonomy when they 'sign up' for certain roles. However, their rights still require attention.

which a scientific advance is weaponised, the victims are not participating in an attack voluntarily (obviously). We tend to think that a benign application of biotechnology, such as adding vitamins to commercially-produced bread, is not harmful. Argument has been made in favour of such activities on grounds of public health, for example. Few would argue against this at first. However, we only need to look at the inclusion of E-numbers (natural and artificial preservatives) in foods to see that additives intended to preserve foods were also shown later to be associated with health problems in children and adults.[8]

We are already familiar with voluntary participation being mediated regularly on behalf of others. For example, in the case of children or of adults of limited capacity in self-determination, others will be required to make some or all decisions for them. Usually, research requiring access to children or to limited-capacity adults (there are many different types of incapacity) will require full ethical approval, precisely because of the lack of voluntary decision-making by all or some of the participants. This is an accepted method of dealing with participation in society today, with various cultural variations. However, as I have mentioned in an earlier chapter, ethical approval in itself is not fool-proof; many 'experts' sitting on ethics committees have little or no formal training in ethics, so their decisions may often be coloured by personal values, power relationships, personal relationships, and other non-ethical concerns. In the context of bio-chemical security, scientists and their associate professionals may in fact be faced with the need for taking necessary decisions that impact on the voluntary 'participation' of themselves, their colleagues, and the public, in their work (or that of others). Think of Werner Forssmann and the nurse we considered in Chapter 4. Any embargo or limitation of participation in research or in its outcomes may be imposed or required at the level of the individual, a group or a population.

During social science research projects, it is vital that people get involved with the research voluntarily, meaning that they know what they are 'getting into'. In the dual use context, this particularly applies

---

[8] See, for example, 'Additives or E numbers'. Available on: https://www.food.gov. uk/science/additives. Accessed 8/12/15.

to all the people *working on the research itself*, directly or indirectly: primary researchers, assistants, technicians, cleaners, administrators, and so on. Does everyone know what they are working on? Obviously some of the support staff do not need to know or understand all of the technical aspects of the work, but consideration should be given to the degree of information they need and in what format it should be provided. Does everyone appreciate (even in a non-technical manner) the possible dual uses of the research? Does everyone have a chance to contribute to risk management activities? Are the voices of all staff heard? Crucially, is the risk management training comprehensive, appropriate and undertaken by all? Is it reviewed and updated regularly? Is it a cast-iron rule that staff cannot start work in the lab, or parts of it, unless and until they have successfully completed appropriate risk management training? If not, why not?

In social science research it is also usually important to make it clear to the participants that they can leave a project whenever they wish. Instructions or guidance should usually be provided showing participants *how* they may leave. In a similar way, when looking for dual use problems in a research project, it is important to identify 'escape routes' for participants — see the list of staff above — and ultimately of course, the public who may be adversely affected by the implementation of any identified dual use. This may be achieved by restricting access to some processes, findings or knowledge; or it may be built in to some technology so that it cannot be reused, for example. We already do this with the implementation of biosafety protocols, but we also need to extend this now to bio-chemical security protocols.

An end to voluntary or involuntary participation may not be possible in some situations, leaving the scientist in a dilemma about how to proceed; in such cases it is likely that only outside intervention could assist the scientist, which then raises its own ethical questions. An example of this would be in a situation where the scientist needs a certain number of staff to carry out a project, or certain materials to work with. Managers and finance directors may not appreciate the requirements of voluntary participation and autonomy; but if people are coerced (however it is done) into working on projects they fundamentally disagree with, trouble can only ensue.

Of course, none of this says anything about *motivation*. Sadly, we may find people involved in all levels of science today whose aim is to misuse that science in the future. Put bluntly, there is little we can do about this other than being aware of it and generally raising awareness of this likelihood amongst the science community. Would you feel comfortable 'vetting' your students or staff before taking them on? What would you be looking for? How would you know you were looking for the 'right' signs? This is where we need to work alongside other agencies and levels of management to agree on appropriate actions (if any) to take to minimise the risks of the misuse of science in terms of our *personnel*. In some countries, the various branches of the military employ Personnel Reliability Programs to vet and monitor staff whose work relates to nuclear weapons and biological and chemical agents.[9] Whether you think this is a 'good thing' or not, I suggest that something like this may well be the pattern of things to come following a major incident involving a civilian lab and a biochemical weapons episode, however indirectly.[10] Governments wishing to be re-elected are more likely to accommodate the concerns of the voting public than uneasy scientists wishing to maintain their traditional 'freedoms'.

Social science approaches to voluntary participation in research

I have included here some further discussion on issues of voluntary participation. These will enable you to see how the principle is applied

---

[9] See, for example, Department of Defense Manual NUMBER 5210.42 January 13, 2015 *Nuclear Weapons Personnel Reliability Program*. Available on: http://www.dtic. mil/whs/directives/corres/pdf/521042m.pdf. Accessed 8/12/15; Department of Defense Instruction NUMBER 5210.65 March 12, 2007 *Minimum Security Standards for Safeguarding Chemical Agents*. Available on: http://www.dtic.mil/whs/directives/ corres/pdf/521065p.pdf. Accessed 8/12/15; *Report of the Defense Science Board Task Force on the Department of Defense Biological Safety and Security Program*, 2009. Available on: http://www.acq.osd.mil/dsb/reports/ADA499977.pdf. Accessed 8/12/15.

[10] Such an incident could be a terrorist event or a state-level activity. While terrorism is often seen as the bigger risk by the public, it is arguably state-sponsored programmes which are the greater threat; think of accounts of the use of chemical weapons in the civil war in Syria.

in the context of general research, without specific consideration of dual use challenges. While reading these sections, and in subsequent sections on the other standard ethical principles, bear in mind in all the examples I give showing how dual use activities could be hidden within superficially benign research, and consider how you may mitigate the risks of this. Consider also how you may adapt these examples into your science practice.

Does advertising for recruits compromise voluntariness?

In some forms of research it is acceptable to advertise in some way for research participants. This may be through emails, posters, letters, and so on. Because participants are *still* self-selecting and join up of their own free will, voluntariness is not compromised. For example, a student, trainee or staff member may advertise for university participants to use a new anti-bacterial soap, and may recruit students or staff who agree to come to a designated site for hand-washing with the experimental soap and a control soap.

However, if pressure, or even implied pressure, were applied to students and staff to join the project, then voluntariness may be compromised. For example, if the soap experiment were to be carried out by a tutor, s/he would be acting unethically if s/he told the class that they *should* join the study. The students may feel coerced even if the tutor has not asked them — care needs to be taken to avoid misunderstandings in power relationship scenarios.

Likewise, a questionnaire attached to an email, or posted out, will in effect be a self-selecting voluntary process as the recipient can choose to ignore it or to fill it in and return it. However, those in power relationships must be aware, in cases like this, that would-be participants should not be made to feel coerced into participation. If the tutor asking for volunteers is also going to be marking assignments later, is this a 'free' choice for the students?

In the science setting, all of this can be related to the selection of staff, students and others who engage with the laboratory willingly. The question is, do they all know what they are about to become involved with? What procedures do you have in place to *inform* people of what the lab is involved with? What level of information do you think is appropriate? Do the people involved agree with you that this *is* an appropriate level of information? Risks of participation (or employment) must be dealt with in the consent process, but still come under the 'voluntary participation' banner.

Can I offer incentives to participate?

This depends on many issues, including where the research is taking place; the population being sampled or reviewed; the nature and aims of the research; the culture in which the research is taking place; the ideals of stakeholders (including funders), and so on. There is no simple answer to this question, so take advice. In the case of educational labs, or commercial labs with promotional prospects as a key motivator of staff, incentives can be problematic if they could be seen to compromise the values of staff in getting involved in something they don't fundamentally agree with. In my opinion, the use of direct or indirect incentives in a lab context are not to be recommended. This could include incentives such as promotion, a pay rise 'if you could see your way to helping me with this', more access to funding, the possibility of becoming a member of a prestigious team, or even something as mundane (but socially important) as getting a space on the 'best' bench in the lab (before you laugh in derision, talk to your juniors).

What about covert research?

Sometimes in social science research it is important that participants do *not* know that they are being observed in some way for data collection. This is a potentially contentious issue because it can compromise voluntary participation. Psychological experiments often rely on some sort of covert observation or deception in the data collection process, as do any research projects relying on observation techniques, or projects which ostensibly say they are looking for evidence of "*x*" when in fact they are looking for evidence of "*y*." To be open about the true purpose of the research from the start would impact adversely on the data collection process and make the project useless or invalid.

In such cases, formal advice should be sought from the relevant ethics approval panel to ascertain if a full ethics approval application needs to be made. It is possible that full approval will be required, because of the potential deception involved and the lack of voluntary participation in an informed manner. However, this may not always be the case.

In the case of the scientist in the lab, 'secret' or 'hidden' research is not a recipe for transparent and open ethical behaviour. We all know that classified research goes on in certain labs, but that is part of our national security framework. To have 'secret' work going on in a lab that colleagues are not aware of, is problematic because it denies those who are not informed of it of their choice to expose themselves to any adverse outcomes of their involuntary association with it. If the lab 'goes down' as a result of a media exposé, all the staff will be tainted by association, not just those who were involved in the 'secret' work. This is an ethical issue as well as a professional issue.

What if participants want to leave the study?

It is vital that research participants be made aware, at the beginning, that they can leave your project *at any time* and that no coercion will be used to keep them 'on board'. This can be dealt with in the consent process. Would-be participants should be given explicit information telling them *how they can leave* the study. This could be as simple as refusing to talk and answer questions in a 'passer-by' survey, or more complex when a participant who is part way through a longitudinal series of interviews or observation sessions chooses to leave. You must give clear information to would-be participants showing them ways to contact you to let you know they no longer want to be part of the research.

It is also useful to researchers to bear in mind that if a participant leaves the project, he can also request that his data should be also removed from the project's records. He may, of course, be happy to leave his data behind, but he must be given the choice. If this happens, researchers must just 'live with it,' no matter how painful it may be to the research.

In the science lab setting, staff may leave; what information and knowledge are they taking with them? What materials are they taking with them? Are you sure that all your inventory processes are robust? Support staff may be in possession of far more information about what is going on in the lab than is recognised. Even outsiders will know more than you think. I have had some very interesting conversations with taxi drivers taking me to and from Porton Down[11] in the UK. While they may not be fully informed, they have certainly passed on to me information that would be of interest to people motivated to misuse science.

The Ethics Toolkit: Implementing Voluntary Participation in Science

- Do all my would-be participants or research/work colleagues know that they do not have to take part in my research if it has dual use potential?
- How can I assure myself that this is the case?
- Do all participants or research/work colleagues understand what the research/work is about and its potential implications?
- What do I need to do about this?
- Are any benefits or incentives offered? If so, are they agreed (by whom?) as appropriate?
- Am I exerting undue pressure on would-be participants or research/work colleagues?
- Is there any power relationship here? (Usually!)
- If so, how do I minimise or remove the potential effects of this?
- Do those giving consent on behalf of others understand the implications of participation?
- Is the capacity of third party consent-givers acceptable and is their consent-giving relationship to the participant recognised as legitimate? (for example, are you assuming that all of your support staff and so on are 'happy' with what you are working on? How would you know that?)

---

[11] Porton Down is a military science site belonging to the UK Government. The Defence Science and Technology Laboratory (DSTL), supporting the UK Ministry of Defence, is located there along with other facilities.

- If I am undertaking some covert or indirect research, have I sought and found suitable advice and guidance from more experienced colleagues if necessary? (This is a major issue in potential dual use contexts — and covert work should be avoided in civilian labs wherever possible.)
- Do I need to gain formal approval for my research from an ethics panel or review board? (How do you know they really understand the issues?)
- Is my research governed by any external body that needs to be consulted? (If in doubt, consult — it's too late when the pathogen is breeding happily among the local bird/animal/human population.)
- *Am* I seeking consent from reliable and valid would-be participants or research colleagues?
- *Should* I be seeking such consent?

How would you answer these questions in a dual-use risk assessment?

Dual Use Bio-chemical Security Ethics: Voluntary Participation

Intervention Point 1:

You are working on benign research that may have dual use potential. You need to look out for this and take steps to minimise the risk of dual use being implemented by others. *How can voluntary participation be used as a tool in promoting this?*

Intervention Point 2:

You do not know that you are in fact working on projects that are being used for dual use purposes. *How can voluntary participation be used to avoid this situation occurring?*

Intervention Point 3:

You discover later that you are or were working on projects that are/ were being used for dual use purposes. *How can voluntary participation be used to avoid this or to mitigate the ill-effects of this knowledge causing harm to the individual or others?*

Intervention Point 4:

You are pressured to engage in dual use activities. *How can voluntary participation be used to avoid this? What other employment systems could be used to support workers, give advice and provide alternatives? Is any of this possible?*

If you are a scientist, how do you react to the Intervention Points *in terms of voluntary participation?* What can you do *through voluntary participation*, to minimise the risks of:

- Being involved, or allowing others to be involved, in dual use work?
- Other people misusing your work?
- Other people engaging in their own dual use work?

Think about what level and type of voluntary participation, if any, needs to be obtained for:

- Yourself,
- Your colleagues,
- Associates of the work,
- End-users of the work/research,
- The public.

At which stage of research should your risk-minimising activities through voluntary participation be implemented:

- Research design stage?
- Data generation stage?
- Data analysis stage?
- Publication stage?
- Post-publication stage?
- Later? And where?

Who and what require protection from hostile misuse of your work? What can you realistically do about it?

Deontology and Teleology of Voluntary Participation

Look at the continuum below and consider your responses to it.

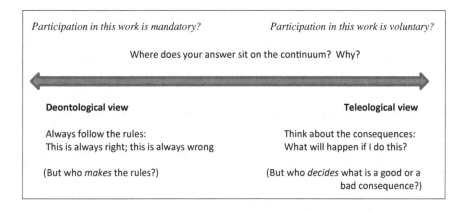

What considerations do you need to take into account when deciding where on the continuum your proposed answer lies?

Consent

Consent issues are related to voluntary participation and the giving out of sufficient information to would-be participants or colleagues to enable them to make a reasonably-informed decision about self-determined participation, bearing in mind all the relevant facts.

Clearly, in relation to dual use bio-chemical security, this involves the consent of those engaged in the work and those who may be associated with the work (including preliminary work upstream and subsequent work downstream). Basically, if you are engaged in work that could be, or is, being used for dual use purposes, are you aware of that fact and have you consented in some form to this? Conversely if you are involved in dual use work unknowingly, where do your responsibilities lie in relation to what you *should* know about the work in which you are engaged? And what about your responsibilities for *finding out* about the potential for dual use of the work in which you are engaged?

Consent in terms of population-level agreement is not an issue here *per se*, if we assume that it is a given that no population would consent to being subject to a dual use attack of any sort or to supporting the development of biochemical weapons. This principle is more applicable in the dual use bio-chemical security context to individuals and groups of scientists working either directly or indirectly on research with dual use potential or actual dual use application. Is ignorance of purpose an effective defence for a scientist? Another consent issue relates to the scientist *consenting* to others having access to his work. Allowing or requiring others to get involved in work that has potential dual use applications amounts to allowing or requiring them to be involved in something that has legal implications under the BTWC and the CWC. While we don't think of this on a daily basis, we need now need to do so. Ignorance is not a defence.

One of the greatest popular misconceptions about consent is the belief that giving consent is a one-off event. It is not. Giving consent to participate or engage in research is an *ongoing process of agreement* to take part that lasts and remains open to challenge until the publication stage, possibly even later than that if the researcher wishes to re-use data. Once given, consent, or at least the effects of it, may remain in place for ever. If you consented to be part of a research team that produced some major dual use technology, even if you were only involved in benign work with it, *you are still associated with the work even when it is misused*. What does this mean to you in practice? Would you do anything differently if you knew in advance what might happen in the future?

Consent may, of course, be withdrawn, on the part of the researcher or the participant at any time. In the dual use context, this could involve the PI, research assistants, technicians, or even administrative personnel. Anyone who is associated with research of concern ought to be in a position where they can withdraw their consent to continue in that association. Clearly this is tightly bound up with voluntary participation and the two principles cannot easily be unwound. The problem is that not everyone knows what it is that they are associated with. Should they? There are no easy answers to this question. Whatever your view, it is clear that scientists and associates need to consider how they may manage consent issues within their teams and beyond.

This brings us on to the people who do not belong to the facility but who engage with it regularly. It is easy to think that consent or any ethical issue ends at the laboratory door or with the funding body or the journal editor. But it can extend over long distances in time and space. A scientist may be satisfied that he has managed all ethical issues appropriately in his sensitive work, but what about the consent or voluntary participation of the truck driver who transports the physical results of the work from the laboratory to the wholesale outlet or another laboratory? I have already mentioned taxi drivers above. Another everyday example of this would be to consider the role of a pharmacist being required to dispense contraception or the 'morning-after pill' when it is against his religious beliefs to do so. This is not typically considered to be a dual use issue, but contraceptive medication has been developed from earlier work on hormones and human physiology that has also led to the development of medications to help with non-contraceptive related medications. Who would normally think of that? Where does consent play a role in this?

Consent also carries with it an implication (on the part of the participant and of future audiences) of the researcher's skill and capacity to actually carry out the research properly. Put bluntly, is the scientist suitably qualified to be doing what he is doing? This is important because the researcher, if not suitably qualified for the research involved, may cause unintended and potentially serious harm (see the case, in Chapter 8, of smallpox at the University of Birmingham in 1978, where an inadequately qualified PhD student was left to work alone). There is, of course, always the possibility that a scientist may accidentally enable some dual use potential in his work and not even recognise it at the time. Even the action of penicillin was discovered by accident. In today's litigious world, participants discovering the origins of such accidental dual use later may well have a case against the researcher and others involved in the research process. How would you manage this situation?

In most research practice there are three different *levels* of consent, which are defined by the amount of information given out to the participant(s):

- partially informed consent,
- reasonably informed consent,
- fully informed consent.

and two different *forms* of consent, which are defined by the way in which consent is acquired and/or recorded:

- written consent,
- oral consent (which may be off-the-cuff and not recorded, or recorded by video/audio prior to a data generation session).

Consent may be assumed or specifically addressed. These depend on the nature of the research, the participants and the context in which the research is carried out. You need to be familiar with the uses and limitations of all of these forms of consent and understand why they are important. In the context of dual use bio-chemical security, we need to consider the following groups and types of consent to be implemented.

Whose Consent?

Research staff — Principal investigator, co-investigators, research assistants, technicians, administrative personnel; these people usually know what they are involved in (or should do, partially at least), but must still give some form of consent to that involvement;

Participants — individuals, groups, populations; these people must know something about what they are involved in, or should do;

Associated people — laboratory cleaners (what are they being required to clean up?), lab visitors, students, colleagues, equipment installers, others in the building, area or beyond who come into contact with the work indirectly at the time or later, and so on; these people may not know anything about what they are exposed to — should they?

Clearly a lot of concerns are addressed through existing biosafety procedures, but is it appropriate to assume the consent to involvement of *everyone* who comes into contact with the work even indirectly? No easy answers of course, but we do need to think about these issues instead of making assumptions about the opinions of others based on our own knowledge, background, and attitudes.

When considering consent, we obviously need to consider what it is that we asking people to *consent to*. This may be some form of simple participation on the part of research staff or of participants. Their role may be small or large, of short or long duration, and it may occur at any stage of the research process. This needs to be clarified in the mind of the chief and his/her staff and of those taking part in any way. See below for further discussion (in general terms) of the different types of consent as outlined above. Then try to think about how these may be or ought to be managed in the dual use bio-chemical security context.

In practice, it is likely that some people involved with the work will know *everything*, some will know *a little*, and some will *know nothing* about the potential or actual dual use of the work they have contact with. We need to consider, in each situation, if this is right (this is not an easy definition to achieve!), appropriate and whether it can be defended ethically. Decisions should be discussed and recorded in writing (minutes, notes, policies, and so on). Each situation will be subject to its own set of circumstances and challenges, as well as in the potential scope for harm, numbers of people potentially affected, security issues and need-to-know status. The big question is, where does the power lie and who can influence this?

Social Science Approaches to Consent

I have included here some further illustration of issues around consent. These will enable you to see how the principle is applied in the context of general research, without specific consideration of dual use challenges. While reading these sections, and in subsequent sections on the other standard ethical principles, bear in mind in all the examples I give how dual use activities could be hidden within superficially benign research, and consider how you may mitigate the risks of this. Consider also how you may adapt these examples into your own science practice.

Do I need to obtain consent?

Usually, yes. But it need not be onerous or difficult, or in writing. You are intruding in some way into the lives of your participants or colleagues. You are either asking them to do you a favour by letting you ask them questions, do something to them, watch them, listen to them, follow them or whatever, or you are requiring that they as part of your team engage in certain activities. You may be asking them for some very personal information. On the other hand, you may simply be asking them what they think of the University website or something similar that should be relatively innocuous. In the case of staff, you may be requiring that they perform a different task or learn a new role. Either way, it is a matter of courtesy at the very least, and of voluntary participation, to ensure that the participant or colleague understands sufficiently to make an informed decision about joining a study or continuing in the job.

Consent may simply be sought in conversation by the individual agreeing to talk to you or to give you some information. You do not need to obtain written (i.e. signed) consent for every research intervention. In the case of taped interviews, it is acceptable to cover consent at the beginning of the interview, and the recorded word is deemed as consent (and it would be later transcribed into writing anyway). In some cases, such as in the return of a questionnaire, it is possible to assume consent by the participant's decision to return the questionnaire. If they do not 'consent' they can ignore it, and you.

However, some authorities do not accept the notion of assumed consent. This is often based on concerns around power relationships. For example, as mentioned above, if you as a course leader choose to send out a questionnaire to your classes, students may feel obliged to return it for fear of you 'marking them down' on their assignments. Such power relationships may indeed negate assumed consent. Think carefully about this. Cultural issues come into play here as well. For some non-western would-be participants, it is difficult to refuse an authority figure's request (even if you don't think you are an authority figure).

There are also types of research which we as researchers may consider intrusive or sensitive, but which may be difficult to carry out if written consent is introduced. For example, the very production of an information sheet to sign may put some people off because it immediately looks very 'official' and even frightening to some people. So use your discretion when deciding whether or not to use written consent. Take advice from more experienced colleagues, the ethics approval panels, or a professional association. Do not leave yourself exposed by making a decision based solely on your own understanding and experience until you are experienced enough to do so. Even then, it is best to share the responsibility of getting consent and deciding how this should be done. Some disciplines or subject areas expect written consent to be used and to not use it would call into question the status or validity of the research in the eyes of the participants (this often happens in business and management contexts).

Some research is intrusive and will probably need consideration of written consent. For example, if you are asking people to participate in interviews that may cover sensitive information about work, health, religion, politics or any other sensitive subject, then you ought to consider written consent. This can be done quite often in the form of an information sheet, in which you tell the would-be participant about the research and why and how they can help you. They may simply be invited to sign it after discussion and questions with you. A copy would be given to the participant as well as a copy kept by you as the researcher.

Such an information sheet should also include in it a paragraph about voluntary participation and tell the participant that they can leave at any time, with or without their data. It should also tell the participant how they can contact you (but probably not your private numbers or address, for your own security) and crucially, *how* they can leave the project (phone, letter, telling you or telling a designated independent person, or simply not responding to a subsequent request for access, for example).

You should also include details of the level of privacy that you are offering and how you will manage this. If you are going to see the participant on more than one occasion, for example to carry out repeat interviews or other data collection, then consent should be revisited *at each event or at least at regular intervals.* You do not need to re-sign papers, but the participant should be asked prior to, or at, each subsequent event or at some other time, if they wish to continue in the research. This means that consent is *ongoing.*

Do I need to give out full information about the research?

Not always, no. In some cases, to give out too much information in the consent process would negate the data generation process. However, a balance must be struck between too little and too much information. You *may* be perfectly happy to give full information about the project (what it is for, what you need from participants, how long it will last, where it will be published, how it will be used, etc.). This may be referred to as *fully informed consent.*

But, you may need to only give out basic or minimal information to would-be participants in order to protect the integrity of your data generation process. If you give out too much detail, it may result in participants telling you what they think you want to hear, or in intentionally deceiving you for other reasons (often without malicious intent). Such consent is referred to as *partially informed consent.* This does not need to involve deception on your part. It may simply be an issue of managing the research design by allowing you to carry out data collection in a difficult or sensitive area.

Many researchers choose to use *reasonably informed consent.* This means that you, as the researcher, decide just how much information is necessary to give out in order to recruit and retain participants. Obviously you need to determine the risks your research poses and inform participants in a reasonable manner as to what the risks are. You may need help and advice in deciding this, which is recommended. In general, if there are significant risks to participants, approval should be sought in principle.

Arguably, most research participants do not wish to know every aspect of the research process and design. Many people feel flattered to have been asked to participate in your research, but this in turn produces something of a power relationship, of which you should be aware. Others feel coerced into research participation (see below on third party consent), but you must respect the rights of participants to know what they are involved in. Issues around risk have considerable overlap with the 'No Harm' principle, and with Beneficence too (see Chapter 6).

Consent given by a third party

In some cases it is necessary or customary for a third party to give consent on behalf a research participant or group of participants. This may be in the case of children or of limited-capacity adults. In such cases full ethical approval is likely to focus significantly on the consent process and how you handle it.

In other cases, certain cultural factors may come into play. Certain scenarios that would not be considered the norm in the West may arise, but you may have to accommodate them. This may include the giving of money or goods or other benefits in kind, as a precondition to consent, that would not be the norm in the West. In some countries and cultures, it is considered usual to ask a husband's consent in order to talk to a wife. The husband may insist on being present at the interview or data collection. This scenario could of course result in a wife not giving you her true opinion, as her husband is listening. You may have to respect this and accommodate it in your research plans. In other cases, it is usual for a village elder to give blanket consent to talk to an entire village, when half the people may not want to talk to you. Gifts may be sought before consent is given. Employers may wish to give consent on the part of their employees. Youth group leaders may try to give consent on behalf of their members. It is up to you to decide how to manage this, under advice. It may be advisable, for example, if you can speak to the 'consented' participant alone to get their consent or not orally, and proceed accordingly, but if you cannot work around this, then your research design may have to be altered. Seek guidance from more experienced colleagues and/ or the ethics panels.

Consent given by other means

Some research does not need to be subject to direct consent by a person as a participant. For example, research on tissue held in tissue banks or other storage facilities has already been subject to some form of consent at an earlier stage, usually during the donation/acquisition stage. Researchers in such cases should act in accordance with the regulations governing their use of such tissues or other materials. However, you need to be aware that you are relying on the consent procedures taken 'upstream' from you, and their policies and behaviour may turn out to have been problematic. Where does that leave you 'downstream' of their actions?

These regulations may be derived from national legislation, professional associations, or a university, for example. It is the researcher's duty to familiarise him/herself with the regulations and how to abide by them. However, it should be borne in mind that consent *has been* a feature of the research at some point in the past in order for the material to be held in its current location. Researchers should be aware that such previously-awarded consent (which may have been assumed) may be challenged in the future and this may cause problems to the researcher and his/her work then. The prime example of this is the former NHS practice of holding tissues without consent for the purposes of research, teaching or even display in museums.[12] In decades past, it was normal practice for clinicians to retain organs or tissues for research or teaching, without informing the patient. There was a lot to be said for this as such use had positive outcomes for medical students and for the benefit of future patients. However, as human rights awareness has increased, many people are keen to retain their

---

[12] Prior to the Human Tissue Act in the UK. See the Alder Hey organ retention scandal elsewhere in this book.

rights over their body parts or those of their dead relatives. As society changes, so do ideas of ownership, human rights and consent.

It is always a good idea to provide the name and email/phone number/business address of a contact person so that participants have another individual to go to if they have any questions about the work. These contact details may be given out on a consent form, in a letter, or on an information sheet. It should be borne in mind that some participants may have questions that arise after some period of time has passed by, so the named person should be someone who is likely to be available for a considerable period. This is also useful because it allows participants to ask questions or to discuss issues that may arise out of their participation.

Summary of Consent Issues

Consent is a type of contract. It implies your capability and qualifications to do the job. Many consent issues can be dealt with orally or by giving out information to would-be participants and having a signature agreement. This does not mean that consent should not be revisited during the participation and/or at subsequent data collection times. Much scientific research involves work on materials that have been subject to consent before they reach the actual researcher. You can cover information about risk and future use of the research, as well as issues of voluntary participation in a simple information sheet if this will help participants. Take your cue from other researchers in the field, and be aware of current legislation, professional guidelines, university or other relevant requirements, and of course, common sense.

In the science setting, consent is all about staff recruitment, retention, management and open and transparent communications. It is often, if it is ever actually considered, assumed that agreement to work in a facility constitutes consent to whatever is being done at that facility. This is probably not a safe assumption to make. Only you can decide what is appropriate for your facility, but you do need to think about how you can transpose these social science examples, where necessary, onto the working arrangements in your facility. Consent is a major ethical issue that is closely related to voluntary participation.

Used wisely, it can help you to avoid a lot of trouble, for you and for the whole facility. If you have staff who do not personally agree with what is being done in the facility, then you have a problem. It is better to recognise this sooner rather than later, when it may be too late.

The Ethics Toolkit: Implementing Consent in Science

- Do I need to obtain consent? [From whom? For what? When? How often?]
- Do I need to obtain written or simply oral consent?
- How do I record this?
- Does the participant/research/work colleague get a copy of the signed consent?
- What information do I need to give out in the consent process?
- Am I assuming consent from participants/research/work colleagues when it should be formally addressed?
- Am I satisfied that previously-obtained consent was obtained appropriately? If it was not, what does this mean for my work now?
- Have my participants/research/work colleagues been given information about how they can leave the study/work/project?
- How often, and when, will I revisit consent with my participants/research/work colleagues?
- Where and how will I store consent records?
- Have I considered all the issues that I can cover in the consent process?

Dual Use Bio-chemical Security Ethics: Consent to Involvement

Intervention Point 1:

You are working on benign research that may have dual use potential. You need to look out for this and take steps to minimise the risk of dual use being implemented by others. *How can consent be a tool in supporting this aim?*

Intervention Point 2:

You do not know that you are working on projects that are being used for dual use purposes; *how can consent be used to avoid this situation occurring and/or to protect the scientist and others?*

Intervention Point 3:

You discover later that you were working on projects that were being used for dual use purposes; *how can the consent process be used to avoid this or to mitigate the ill-effects of this knowledge causing harm to the individual?*

Intervention Point 4:

You are being pressured to engage in dual use activities. *How can the consent process be used to avoid this? What other employment systems could be used to support workers, give advice and provide alternatives? Is any of this possible?*

If you are a scientist, how do you react to the Intervention Points *in terms of consent?* What can you do *through consent,* to minimise the risks of:

- Being involved, or allowing others to be involved, in dual use work?
- Other people misusing your work?
- Other people engaging in their own dual use work?

Think about what level and type of consent, if any, needs to be obtained for:

- Yourself,
- Your colleagues,
- Associates of the work,
- End-users of the work/research,
- The public.

At which stage of research should your risk-minimising activities through consent be implemented:

- Research design stage?
- Data generation stage?
- Data analysis stage?
- Publication stage?
- Post-publication stage?
- Later? And where?

Who and what require protection from hostile misuse? What can you realistically do about it?

Deontology and Teleology of Consent

Look at the continuum below and consider your responses to it.

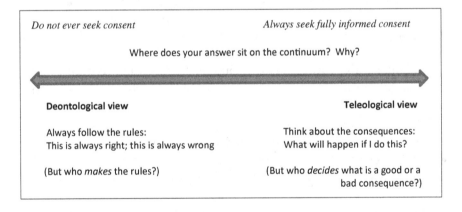

What considerations do you need to take into account when deciding where on the continuum your proposed answer lies?

Privacy

In the dual use bio-chemical security context, privacy takes on a slightly different meaning from what is normally understood by the term in social science research. There is more about the general social science (with science overlaps) approach below and a lot of it applies in our context, but let us think for a few minutes about what privacy, anonymity and confidentiality mean in terms of dual use bio-chemical security.

The issue here is one of secrecy or security, indicating restricted, controlled or filtered access to certain information, processes, things, places or people. We can summarise the possible fields in which we might need to address this here:

- Security of materials, their availability and transport;
- Security of data and processes, physically, electronically, and intellectually;
- Security of access to the workplace or storage place, physically, and electronically;
- Privacy/security of individuals involved in the work (for their own protection, for the protection of the work, or for political, economic, religious or other reasons);
- Privacy/security of others (those with indirect contact, the general public).

As with all ethical issues, there are power relationships active here. There is clear overlap with consent issues and with voluntary participation. As we have discussed above, even in entirely benign and peaceful work, not everyone will know what it is that they are involved in; there are different levels of security to be implemented and maintained depending on the potential or actual dual use risks or current application of the work in question. Whereas in everyday social science research, privacy, in the form of anonymity and confidentiality, is often offered to research participants to protect them, to allow them to provide sensitive information, and to gain access to them, in the dual use bio-chemical security context, privacy and security usually go hand in hand in a more literal sense.

The ultimate security here is of course that of the public — our duty to protect the public from the hostile misuse — dual use — of biochemical technologies or other scientific applications that may be amenable to hostile dual use. All other privacy and security arrangements are implemented with this final goal in mind.

This is covered in more detail below, but in the social science context, anonymity means that an individual's name or other identifier (you can be identified by more than your name) is not recorded anywhere during the research process (except on a secure key list with an associated code identifier, perhaps), not even on research notes or forms and it is *never* published in reports or accounts of the work. Confidentiality means that the individual's name or other identifier *is* recorded on notes and forms and other records, but it is still never mentioned in publication. Anonymity is therefore the more robust of the two privacy methods because even if unauthorised access to the records is achieved, no identifications can be made.

In the dual use bio-chemical security context, we need to consider these two methods for scientific staff and associates as well as for any participants, and possibly even for end-users of the work if there is a very high risk of misuse of the work if any publicity should occur. How might this be managed, especially in terms of recruitment, working practices and employment law? What about voluntary participation and the consent process? Who will manage the privacy system? What are the boundaries of it in time and space? Over what length of time will your privacy and/or security arrangements operate? How will you manage long term privacy or security policies? Is any of this process governed by law or codes of practice?

A major question that arises out of the privacy/security issue is that of accountability. What implications for accountability arise if privacy/security is high? If inadequate consideration and planning is not carried out, it is likely that as security levels go up, clarity of lines of management, openness of discussion about the work and problem-solving techniques and opportunities will all be increasingly obscured to some degree. These are highly likely to lead to a loss of accountability and this may in turn lead to activities being sanctioned without full information being available through the normal research/work monitoring channels. Is this acceptable? To whom? Again, there are no easy answers here, but we do need to consider these issues if we are to engage effectively with dual use bio-chemical security activities in terms of privacy and security.

Social Science Approaches to Privacy

Privacy of the research participant and of the researcher is an ethical issue that can easily be overlooked. It can refer to physical and social privacy and to personal, emotional, and knowledge-related privacy. Consider how you may adapt these examples into your own practice of science.

Practical privacy and safety

Social scientists have to think carefully before giving out their personal phone number or home address to research participants or even some co-workers. If it is necessary, then they must have some system of security and back-up in place. Calling-in procedures are often utilised, in which staff check in with colleagues prior to and after interviews or visits that are potentially hazardous. Clear communications are agreed in advance about where staff will meet with clients or participants, when, for how long and so on. It is not always appropriate to contact research participants at home, at work or in other ways such as email or telephone calls. Their participation in research may not be open knowledge, often for good reason. This impacts directly on the retention of participants in studies and on doing them no harm. Researchers cannot assume further access to participants if it has been granted once, unless a further arrangement has been made. Meeting in neutral places is often a good idea. All of these considerations enhance not only the privacy of the participant in research, but also the safety and personal privacy of the researcher. These are common sense issues, but easily overlooked.

Research privacy

This is governed by two techniques: anonymity and confidentiality. These protect participants from being identified in publications. In some studies, neither technique is required, as participants do not express a need for privacy, or there is no need for it, or participants don't mind being named in reports or other publications.

The difference between the two techniques is how personal data are recorded, kept, and published. In neither case is the participant identified in publications, but with anonymity, further restrictions of identification are in place. Researchers need to consider who will have or need access to the data; who will be involved in the analysis of the data; where it will be kept securely; what future need for access to participants may occur, and whether or not they wish to re-use the data in the future for other studies.

In order to gather valuable but sensitive data, extreme means may need to be employed to protect participants. For example, in some studies looking at very sensitive issues such as sexual health and practices, data collectors are trained to gather data through completely coded data collection techniques in which the participant is never personally identified, is the only person aware of what the question is and gives an answer by a code to the data collector. This takes a huge amount of management on the part of the researchers.

If researchers do not offer some level of privacy, some people may not agree to be recruited. On the other hand, some sorts of research are deemed so harmless that participants do not ask for any privacy at all. This depends largely on the research, the audience, the potential for harm, and the population from which researchers hope to sample.

Confidentiality

This means that no participant is named or otherwise identified in publications, which includes talks, papers, posters, photographs or any other publically-disseminated material, just as in the anonymity scenario. But, the participants' names and other identifiers are recorded on the data collection sheets, tapes or other records. Researchers would not identify participants even in discussion with colleagues unless the participant agrees to this.

This means that the researcher has access to data collection records with names or other identifiers on them. Such records must be kept in a secured location (locked at the very least, and not on laptops or flash drives unless as a very temporary measure; sending such material by email should also be avoided as researchers cannot guarantee the security of the email server). In the consent process, an agreement may be made about how long records will be kept and how they will be destroyed. Agreement would also need to be sought if others are to be involved in the analysis as this may potentially void confidentiality agreements.

Anonymity

This means that no participant is named or otherwise identified in publications, which includes talks, papers, posters, photographs or any other publically-disseminated material. As well as this public protection, the participants' names and other identifiers are not recorded on the data collection sheets, tapes or other records. Researchers would not identify participants even in discussion with colleagues unless the participant agrees to this.

Each participant is allocated a code or key identifier (A, B, C, or numbers, or another code) by the researcher and this coded identifier is used on the records. Usually a key list, linking the code to the actual name of the participant, is kept by a nominated third party (it could be the research supervisor or a colleague), or at least is kept by the researcher, under lock and key. This would be the only means of linking a participant with his/her data. A key list allows an account of the participants' contact details, which is vital if subsequent interviews are required. Loss of a key list is a serious breach of anonymity. In some cases, a key list is not necessary, as data is collected in a one-off event and only some general identifying information is required about the participant, for example, age, gender, place of birth, and so on. Key lists are often destroyed on completion of a project. As with confidential data, the material, even anonymised, should be protected at all times and not kept on laptops, flash drives or sent by email wherever possible.

Anonymity requires more work and management on the part of the researcher but should guarantee the participant total protection from identification if it is done properly. Researchers should not offer anonymity to their participants if there is any chance that they cannot control the process, for now and as far as they can see into the future.

All of these issues can be dealt with in the consent process. Some would-be participants will not agree to engage with the research until they are guaranteed a level of privacy. Details must be given and agreed in the consent process.

Practical security

Security is an important issue if researchers have offered confidentiality, and an even more important issue of they have offered anonymity (the key list, if they have one). Researchers are responsible for maintaining the level of privacy that they have offered. For confidentiality, they must keep all their data records in a secure, preferably locked, place, as respondents' personal details are recorded. With anonymity, they must make sure that the key list, if they have one, is kept in a secure, locked place, as well as keeping their coded data records in a secure place. In neither case will they give any identifiers out in their publications. This is not always as simple as it seems. For example, a researcher may have given an offer of confidentiality or anonymity, and have indeed not given out any names or addresses in the publications. But, have they been able to avoid publishing any comment or other information that may be recognised by others as only coming from Person A or Person B? Sometimes this is impossible to avoid, in which case, the researcher should not have offered anonymity in the first place. The only solution to this is to either change the research questions so that anonymity is not required, or only recruit people who are prepared to accept confidentiality. Not easy!

As well as keeping their records and key lists under lock and key, it is common today for researchers to keep data on computers or other hardware. Before they do this, they must consider how safe it is to keep data on laptops, flash drives, CDs, or on internet storage services such as iCloud, Dropbox, and so on. Small storage units may be easily lost or stolen and researchers cannot guarantee the security of internet service providers' storage facilities. It is inadvisable to maintain any data in such vulnerable locations. Ethical researchers try to keep the bulk of the data in a more secure location, and develop a habit of downloading working data into a more secure location as soon as they can. In this way, if any hardware is lost, only a small amount of data is lost (which is a potentially major issue anyway), which will involve them in a lot of work and potentially the loss of some of their participants, as well as the difficulties such a loss could incur to their participants (see *No Harm*, in Chapter 6).

Security issues also involve the transfer of data by phone, by email or by other electronic means. In some countries, and according to the requirements of some funding bodies, data cannot be freely transferred overseas or by certain modes of communication. It is your responsibility to get this right. Can researchers guarantee the security of the system they are using? Probably not. Bear this in mind. Any lapse of privacy will rebound on them and may have serious, not to say disastrous consequences for participants. I have heard cases of marriages breaking down (current wife did not know of husband's previous family), revelations of alcoholism in the family (two cousins attended the same research-based self-help group for people with alcohol addiction issues; neither had known that the other was 'an alcoholic') and the identification of previously-unknown paternity in families (DNA tests showed that Daddy could not really be Daddy after all) , resulting from failed anonymity or confidentiality in research. Be aware!

Time

Researchers need to discuss with participants, in the consent process and later, how long they will need to keep the data and identifiers on their records. The project may only require the data to be held for the duration of the study. Other projects require long term retention of the data. All of this needs to be decided in the planning and funding stages, as it is often too late to make changes once participants have been recruited. If it has been agreed to destroy data after the study is complete, then they must be destroyed as agreed by suitable means that will maintain the confidentiality or anonymity that was offered. This would probably mean shredding, but could also involve destruction of tapes, CDs, photographs, or other non-paper records, which may take some work to do appropriately. Researchers also need to think about how they will assure participants that the data have been destroyed.

If the work is likely to be referred to by other researchers in the future, researchers may agree to destroy the key list but maintain the coded records for future use. This could be covered in the consent process as a catch-all at the beginning if necessary. Researchers may wish to use their data again in their own future studies. However, unless this was consented to by participants in the initial study, it may not be possible. In cases like this, researchers need to take advice from their ethics approval panel, or from other relevant authorities on how this may be done if they are unsure. Usually the data are considered the researcher's intellectual property, but institutions may also have an IP claim. Further, there may be other stakeholders who have a say in the retention and future use of confidential or anonymous data. All of this involves ethics and none of it is easy.

Summary of Privacy Issues

If researchers have offered confidentiality or anonymity, they are taking on a major responsibility to the participants. If they fail to maintain these, they may cause significant harm to the participants or to themselves or others. Anonymity requires a good deal of management, as does confidentiality, through security, access and publication issues. Can researchers handle these?

In laboratory work, where researchers study samples without personal identifiers being associated with the material, issues of confidentiality, and anonymity do not always emerge. However, privacy and its implications do need consideration even so, because we do not always know what may emerge in even the most harmless-seeming scenarios. In social science contexts, privacy is always a concern. It relates to the boundaries we set with others around access to them, and to publication of their participation. This is not as straightforward as it sounds.

To take a medical example, for instance, let us say someone has volunteered to donate a blood sample as part of a study into Condition A, which is a socially-sensitive disease such as tuberculosis. He has been guaranteed anonymity and it is on this basis that he has agreed to take part, even though he is only part of the control group. Only his age, nationality, ethnicity, and general medical history have been included in the records. However, the research methodology changes during the project and the team begin to look for further signs in the blood. As a result of this, it becomes apparent that our blood sample donor has the HIV virus. If we have granted him anonymity, will we have his name, address or telephone numbers recorded anywhere?

If we do have this, we can contact him to ask him to go to his doctor. But even this is not as simple as it sounds. What do we say to him when we speak to him? 'We have found the HIV virus in your blood sample and you need to see your doctor'? What about the ethical issues involved in announcing that to someone who is not expecting it? (He has a weak heart and the shock brings on a myocardial infarction). If we simply tell him to go to his doctor without revealing why, we may be 'condemning' him to a period of worry or even panic leading up to the appointment with his doctor, which could take some time. (What if he is already clinically depressed, or suffers from anxiety, or has a relative that has recently died from cancer? He may guess, wrongly, that he has cancer too). He may conclude that if we have not told him what the problem is, then it cannot be too bad; he then decides not to bother going to the doctor. What then? (He is generally in good health so he thinks it will be a fuss about nothing). We know he has the HIV virus but we have not told him. He later develops fill-blown AIDS and dies. Who is responsible then?

If we do not want to reveal the finding to him when we tell him to go to the doctor, and if he asks us why, do we lie to him? Or do we tell him? What if he is now out of the country and his wife answers the phone? She says she will pass a message on to him if we can say what it is about. What now? We can not divulge anything to her because she may not even know that he was involved in the research. There may be a good reason why he has not told her about his involvement. And so on... where does ethical behaviour take us? By this time you are no doubt wishing that you had decided to run a market stall instead of a public health laboratory.

It does not take a rocket scientist to work out the implications of all of this in relation to lab staff and all the associated support staff. Safety is one thing, but security is another. Everyone who works with you in your facility is entitled to both safety and security. Everyone outside your facility, beyond the lab door, is entitled to safety and security. Can you offer these reliably?

The Ethics Toolkit: Implementing Privacy/Security in Science

- Do I need to publically identify participants/research/work colleagues?
- Do I need to offer any privacy to participants/research/work colleagues?
- How can I agree this with them?
- Do research/work colleagues need anonymity or confidentiality in terms of being identified with the research/work?
- How can I agree this with them?
- Are there any known risks or dangers for my colleagues in being associated with the work of our facility?
- Do I need to do anything to mitigate these risks or dangers through privacy mechanisms?
- What are the social and/or economic implications for colleagues in such cases?
- Can I actually manage and deliver what I am offering?
- Who has access to what, and where and when?
- Am I satisfied that there are sufficient and appropriate control measures in place?
- Can I control the security of my privacy measures?
- How can I securely manage the transfer of data from the field to the record store, or from lab to lab?
- Have I considered the possibility of re-use of the data in the future and how this may impact on anonymity and retention of key lists?
- And so on......

Dual Use Bio-chemical Security Ethics: Privacy and Security Issues

Intervention Point 1:

You are working on benign research that may have dual use potential. You need to look out for this and take steps to minimise the risk of dual use being implemented by others. *How can the use of privacy/ security techniques be a tool in supporting this aim? How may such techniques be a tool in enabling hostile dual use to occur?*

Intervention Point 2:

You do not know that you are working on projects that are being used for dual use purposes; *how can privacy/security techniques be used to avoid this situation occurring and/or to protect the scientist and others? How may privacy/security techniques be a tool in enabling hostile dual use to occur?*

Intervention Point 3:

You discover later that you were working on projects that are being used for dual use purposes; *how can privacy/security techniques be used to avoid this or to mitigate the ill-effects of this knowledge causing harm to the individual?*

Intervention Point 4:

You are pressured to engage in dual use activities. *How can privacy/ security techniques be used to avoid this? What other employment or other social systems could be used to support workers, give advice and provide alternatives? Is any of this possible?*

If you are a scientist, how do you react to the Intervention Points *in terms of privacy/security?*

What can you do, through the implementation of privacy and security, to minimise the risks of:

- Being involved, or allowing others to be involved, in dual use work?

- Other people misusing your work?
- Other people engaging in their own dual use work?

Think about what level and type of privacy and security, if any, needs to be obtained for:

- Yourself,
- Your colleagues,
- Associates of the work,
- End-users of the work/research,
- The public.

At which stage of research should your risk-minimising activities through privacy and security be implemented:

- Research design stage?
- Data generation stage?
- Data analysis stage?
- Publication stage?
- Post-publication stage?
- Later? And where?

Who and what require protection from hostile misuse? What can you realistically do about it?

Deontology and Teleology of privacy and security

Look at the continuum below and consider your responses to it.

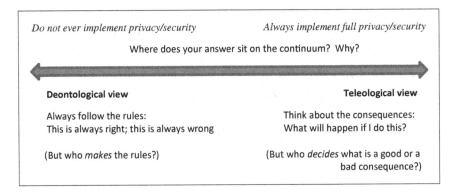

Do not ever implement privacy/security   *Always implement full privacy/security*

Where does your answer sit on the continuum? Why?

Deontological view   **Teleological view**

Always follow the rules: Think about the consequences:
This is always right; this is always wrong   What will happen if I do this?

(But who *makes* the rules?)   (But who *decides* what is a good or a
                                                 bad consequence?)

What considerations do you need to take into account when deciding where on the continuum your proposed answer lies?

That's it for this section on the 'rights' of others — although you are probably thinking that there were an awful lot of 'responsibilities' in this chapter as well. Give yourself time to chew over all of this before going on to the next chapter. That will look at your *explicit* ethical responsibilities towards others.

Chapter 6

The Ethics Toolkit 2: Ethics as Your Responsibilities to Others

The previous chapter considered the *rights* of other people within the framework of bio-chemical security ethics. Let us look now at your direct *responsibilities* under the same ethical framework. Again, there is an amount of overlap that mixes the two concepts of rights and responsibilities.

The No Harm Principle

In the context of bio-chemical security, the principle of 'do no harm' is the basic building block on which all ethical and decision-making processes are founded. It is absolutely obvious (or should be) that the key aim of all bio-chemical security activity is to minimise the risk of harm to the human population, the environment and animals as a result of the hostile dual use of scientific research and advances. This may seem so straightforward that some may argue there is nothing more to be said about this particular principle. There is even an argument to be made that the No Harm Principle should not be an individual principle in itself as it is so foundational to all other ethical principles.

I would argue that it is worth looking at the No Harm Principle as a stand-alone principle in the same way we look at voluntary participation, consent, privacy, and all the remaining principles. To say that we all agree on the No Harm Principle is easy, but that consensus assumes that we can all agree on what a definition of 'harm' actually is. There is usually, or at least potentially, a huge if invisible cultural gulf between any two or more groups of people when it comes to defining harm. There are different attitudes to suffering and hardship when interpreted through the lenses of gender, age, occupation, ethnicity, socio-economic status, geographical location, religious views, educational status, political allegiance, and so on. This is a potential minefield in itself and we could devote entire books to the subject. However, let us just look at some of the possibilities so that we can get started on a better understanding of how to identify harm and from that, move on to thinking about how we can avoid or minimise it.

You will see from the general notes below on the No Harm Principle in social science research that it is probably *never* possible to guarantee that participants in research, or others coming into contact with the effects of it, will be protected from any or all harm. We simply cannot predict or assure a fully 'de-harmed' scenario. This is because any intervention by us, or by someone else using our work, or by intentional or accidental involvement with what we are doing, can potentially cause harm of some sort at any time and in any place from that point on (think about Galston and his fertilizer in Chapter 1).

Are We Ethically Competent?

Here's what we *can* probably say:

We will do our best to protect those involved (define 'involved') or who may suffer (define 'suffer') from potential harm (define harm) as far as we can anticipate (assuming that we recognise something in the first place in order to anticipate it) it in time and space (assuming that we know when and where might the harm 'stop' having an effect), given the information we have at the time of the research (which may be inadequate or not possible to anticipate for good reasons) or work taking place and what we know of those who may be affected (which may be inadequate or incomplete).

Hmm. This is full of holes!

So what can we now say about our ability to implement the No Harm Principle?

- We may not be able to define 'involved' fully;
- We may not recognise potential or actual harm at the time of the work;
- We may not recognise harm after the work is complete;
- We may not appreciate the nature or the extent of the harm itself;
- We may not recognise the extent of the effects of the harm in time or in space;
- We may not know, or be able to know, when the harm will stop having an adverse effect on humans, animals or the environment.

Is Galston's fertilizer springing to mind? In the dual use bio-chemical security context we need to consider harm from at least three perspectives:

- Our own perspective:
  - given the knowledge and skills we bring to the research or work, our *competence* is expected and implied,
  - given that our research participants, funders, employers and others 'behind' and underpinning the work have an *expectation of our capability to carry out the work safely.*

- The perspective of others involved in the research or work:
  - Who also have the same expectations.

- The perspective of those who may be affected by it:
  - who are relying on someone else (i.e. you) to make the relevant and appropriate decisions about their safety and security, whether they know the risks or not.

We can see from this that our competence is a major foundation on which all else is built. Are we ethically competent? How do we know that we are ethically competent?

So we are not only looking at the work itself, but our competence to assess it and act appropriately, from an ethical perspective. This means that we need to be capable of recognising an ethical problem when we see it. Then we need to be able to work out how to address that ethical problem to either eliminate the problem (ethically) or minimise the risk of adverse outcomes (ethically). All of this means *doing* something to openly and transparently reduce risk. Brushing problems under the carpet, or 'kicking things into the long grass' for someone else to deal with later is not an option.

This is no different from the usual social science perspective, but in the case of the bio-chemical security context the *scale* and *severity* of time, space and potential harm that we must consider is usually much wider. A social scientist researching an issue with a small number of individuals may be in a position to potentially affect only his participants, or only one or two small communities of people. A scientist working on research with potential or actual dual use applications has a significant responsibility to identify and mitigate the risk of these being used by others for hostile purposes. The scale of potential damage could be at a population level potentially affecting hundreds of thousands of people or more; even at the local level, the work, if misused, could affect dozens of people. Besides, if you are one of the affected people, you do not really care if it was a 'small' or a 'large' event — it was *significant* to you because you suffered. As scientists, we can be sure that media coverage and possibly lawsuits will follow.

This responsibility is a daunting prospect, and while an individual scientist or associate may not be in a position to identify all risks, or devise and take all the necessary steps to minimise the risks, a group or team of scientists and associates, including those invited to advise from outside the scientific world, would probably be more able to do so — or at least *attempt* to do so. This is, after all, the nature of the ethical responsibility posed by dual use research of concern. Does it not make sense that all science students, and professional scientists, should be educated in applied research ethics at every stage of their careers? If we don't manage to bring this about, as I said in an earlier chapter, someone else will do it for us.

Before looking at the more general No Harm Principle notes below, let us just remind ourselves that there are three perspectives on the application of this principle in the bio-chemical security context:

- The protection of the of the public is as important as the protection of researchers and their associates *plus* indirectly associated workers;
- The protection of the research itself (materials, data, methods, publication and storage, and so on) is crucial;
- The scale and severity of potential harm is potentially far greater than is usual in 'everyday' social research.

Bear in mind that in terms of 'no harm', *participants* can include anyone who is involved in the research, associated with it or with the researchers, and all levels of contact beyond that up to the widest possible population levels.

Social Science Approaches to the No Harm Principle

The No Harm principle is based on the idea that participants should not come to harm by their participation in research. However, while everyone agrees that this is worthy and to be pursued, it is not always easily accommodated.

What is harm? It can be physical, mental, emotional, or it can involve social, employment, political, religious or other sorts of harm. It may even involve loss of life in extreme cases. Not all of these may be foreseen. The key point for you to consider as a scientist is, 'How do I minimise harm to humans, animals, and the environment, both now and after the work is over?'

Risk — to participants, researchers, and others

If there is some obvious risk to the participant, it must be explained at the consent stage — in the bio-chemical security context, this applies to science associates and staff as well. For example, although much previous work has been done to identify side-effects before human trials begin, all human participants in medical trials are warned that they face potentially unforeseen harm. This is covered in the consent process.

There are other sorts of risk, though. As well as risking their health, people may risk their employment, their marriages, their membership of some group or community, their income, their status, their self-esteem and more, by participating in your research or work. Social science research will cause change, often in the participants or their associates. You must consider the risks and address them in the consent process. This is where confidentiality and anonymity often enable participants to join a study. You can see immediately how breach of these guarantees of these can cause serious harm to the participant.

There is also a risk of harm to the researchers and others. For researchers, there may be physical risks, as well as emotional and mental risks. How long should you work in a stressful environment? What are the implications of hearing constant narratives of torture, war crimes, and other disasters for the researcher? What about risk of infections or other health issues? What about the risk of kidnap or being killed in the process of data collection? For others too, there are risks. What about the effects of research participation on the families or colleagues of those who have participated in your research?

If your research causes change (which it always will to some degree — otherwise, why are you doing it?), then who is responsible for the changes and the effects of this? Will your questioning change the views of a wife about her husband? Of a boss about how he manages his staff? Of a village elder about how he treats the vulnerable? You can see that these are all legitimate risks of the research process that could inadvertently cause harm to others on the periphery of the study. Think of this as well: what will be the effect of your research on your own family and colleagues? This is an ethical issue too.

What of researchers working in more liberal countries whose research may well be frowned on in their home country? How will the research affect their lives if they return home? What about their family and friends who are in the home country? These are examples of risk that do not receive much attention, but they are valid and real. In practice, we cannot be responsible for every change and every outcome associated with our research. However, we should be aware of the possibilities, and devise ways to minimise harm, emphasise benefit and add to the experience of our participants, colleagues, and others as best we can.

Working with risk

You may well recruit participants who will take on some risk in the hope that it will help them or others to achieve some personal advantage. Be aware that in such cases, the consent issues of voluntary participation and privacy must be respected. Participants under such conditions may be more likely to wish to leave the study and you must be prepared to accommodate and live with that. You should always be aware that some people may view participation in your research as a means of forwarding their own agenda or getting their own needs met. This may lead to you not accepting them in your project, or in you having to make it clear that you cannot solve all their problems for them. This may result in bad feeling and could cause you further problems, so be aware from the early stages how to avoid this scenario as much as possible.

There are some studies that cannot be completed without causing some harm to participants. For example, let us say that you are gathering data about how counselling can assist victims of war crimes to recover from their traumatic experiences. Just by speaking to you, the participants will relive, to some extent, their traumas. Does this mean that you should not do the research? No, but you should go into it with your eyes open and be prepared to re-design your study to minimise the effects of this trauma, and to provide what help and support you can. In some cases, a study may offer some benefit to one group and withhold it from another.

For example, you may devise a study to test the effectiveness of Process A in helping a group of students to successfully complete their course. You offer the Process A training to one group and do not offer it to the other group. At the end of the course you evaluate the success of both groups. Have you harmed (relatively speaking) the group who did not receive the training? You may decide that in order to ameliorate the effects of missing the Process A training, you will offer it later to the non-trained group, as this will benefit then in future studies. In this way, you have minimised a risk of missing out on something beneficial.

Control of risk

In many scenarios, risk is controlled by legislation. Professional associations or individual institutions may also limit the amount or type of risk to which you, your colleagues or your participants may be exposed. It is your responsibility to acquaint yourself with the requirements of the country, the institution, the community or so on in which you will be working. You may find you are faced with ethical dilemmas because you are working in a country or context which has different notions than you of what is acceptable. There are clashes between what you need to do in the research process and what is considered acceptable in your country or in your university. This requires advice and guidance. You will gain valuable insights into managing risk by talking to colleagues who have more experience (even if it is just how to avoid the messes they got into).

As adverse outcomes of risk are highly likely to result in legal action today, it is worth giving this a lot of consideration. It may be more circumspect to alter your research design if the risks look too difficult or dangerous to manage, for you and for those associated with the study. If in doubt, consult as widely as you can, and seek guidance from your ethics panel.

In the Science Context

In the science context, you can see from all this that your responsibilities to do no harm through your work are wider than you probably thought. We often take it for granted that we will not harm ourselves, but we equally tend to think that 'harm' in the lab is defined only as unwanted exposure to dangerous agents. In fact, harm is a much wider concept and can affect a large number of people indirectly. Obviously, none of us want our work to be misused by ill-intentioned individuals or groups to result in the use of bio-chemical weapons. If you have read earlier chapters, you will now be aware that a biological or a chemical weapon does not just take the form of the 'traditional' versions of these — mustard gas, chlorine gas, weaponized anthrax, and so on. Harmful agents — chemical or biological, added to a public water supply, or to commercial foods, or propelled through air conditioning systems are all relatively easily-achieved 'weapons'. This is why we need to be ever-vigilant in our work. Even an incomplete inventory of lab agents could enable someone to walk out with something that should never leave the lab. Mis-labelled bottles, poor storage techniques, poor waste-disposal techniques and so on could all facilitate the acquisition of a harmful agent by an ill-intentioned person. Massive harm could then be the result.

Think again of the ripple diagram. Once 'it' is out there, what can stop its spread? There are no barriers between the ripples to protect people once an agent has escaped the confines of biosafety and bio-chemical security procedures in and around the lab. This is the potential nature and scope of harm that can be caused by legitimate, peaceful scientific work.

The Ethics Toolkit: Implementing 'Do No Harm' in Science

- In the context of my work, what is harm?
- How would I recognise that?
- Who is it harmful to?
- Is there any reasonably foreseeable risk associated with participation in, or association with, my research/work?
- Who is affected?
- Immediately?
- Soon? (Can you define 'soon'?)
- Later? (Can you define 'later'?)
- Where may the presence of harm become apparent? (You cannot see infection until symptoms appear),
- Can I reasonably reduce the risk of this harm?

- Is there a less harmful way in which I can achieve the desired outcomes from this research question?
- Am I prepared to stop, delay, or otherwise discontinue my work if a certain threshold is reached?
- Who decides what that threshold is, or should be?
- How can I effectively communicate all of this to participants in my work? (Remember that a far wider pool of people 'participate' in your work than you think — your family are exposed to *you*, your colleagues and their families are exposed, all people who visit your facility are exposed, and their families and friends after them — where do the ripples end?)
- How much information about this should I give out and in what form?
- If harm emerges during the project, how can I and participants communicate about it and do something about it?
- What measures can I and my colleagues put in place to minimise risk?
- Can we do anything to provide support for participants or associates who may be affected by harm?
- What *should* we be doing to achieve this?
- Have I considered how to manage any clashes between the values and norms of my country or workplace and those of people from other communities?
- How can we *all* go about agreeing on definitions of harm, and the necessary steps to respond to harm when it occurs?
- Where will all this be codified?
- Who will be responsible for oversight of this?
- How often will it all be reviewed and revised?

This list could go on for ever, but you get the idea.

Dual Use Bio-Chemical Security Ethics: The No Harm Principle

Intervention Point 1:

You are working on benign research that may have dual use potential. You need to look out for this and take steps to minimise the risk of dual use being implemented by others. *How can the application of the No Harm Principle be a tool in supporting this aim? How may the No Harm Principle be a tool in enabling hostile dual use to occur?*

Intervention Point 2:

You do not know that you are working on projects that are being used for dual use purposes. *How can the No Harm Principle be used to avoid this situation occurring and/or to protect the scientist and others? How may the No Harm Principle be a tool in enabling hostile dual use to occur?*

Intervention Point 3:

You discover later that you were, or are, working on projects that are being used for dual use purposes. *How can the No Harm Principle be used to avoid this or to mitigate the ill-effects of this knowledge causing harm to the individual or the wider population and environment?*

Intervention Point 4:

You are pressured to engage in dual use activities. *How can the No Harm Principle be used to avoid this? What other employment or other social systems could be used to support workers, give advice, and provide alternatives? Is any of this possible?*

If you are a scientist, how do you react to the Intervention Points *in terms of the No Harm Principle?*

What can you do through the No Harm Principle, to minimise the risks of:

- Being involved, or allowing others to be involved, in dual use work?
- Other people misusing your work?
- Other people engaging in their own dual use work?

Think about what level and type of 'no harm' needs to be obtained for:

- Yourself,
- Your colleagues,
- Associates of the work,

- End-users of the work/research,
- The public.

At which stage of research should your risk-minimising activities through 'no harm' be implemented:

- Research design stage?
- Data generation stage?
- Data analysis stage?
- Publication stage?
- Post-publication stage?
- Later? And where?

Who and what require protection from hostile use? What can you realistically do about it?

Deontology and Teleology and the No Harm Principle

Look at the Figure below and consider your responses to it.

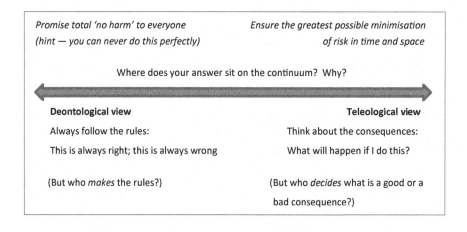

Promise total 'no harm' to everyone   *Ensure the greatest possible minimisation*
(hint — you can never do this perfectly)   *of risk in time and space*

Where does your answer sit on the continuum? Why?

Deontological view   **Teleological view**

Always follow the rules: Think about the consequences:

This is always right; this is always wrong   What will happen if I do this?

(But who *makes* the rules?)   (But who *decides* what is a good or a
                                                        bad consequence?)

What considerations do you need to take into account when deciding where on the continuum your proposed answer lies?

The Beneficence Principle

The principle of beneficence, or of doing good, is based on the idea that participation in research could reasonably be expected to benefit the participant or others either now or in the future. While most research is carried out with a view to benefitting the living (human or other life-forms) or the future living, it is interesting to note that while it is required that we avoid harm, there is little stated requirement to actually *benefit* participants or others in the here and now. Beneficence seems to be simply assumed. More focus is placed on avoiding harm (which is no bad thing). There is a whole discussion to be had about that, but this is not the place for it.

Usually in social science research (with exceptions such as in participatory research and some other methodologies) the benefit to the *researchers* is not such a major or immediate consideration, or much of one, when designing and carrying out research. Most if not all consideration tends to be given to the participants' benefit rather than that of the researcher. This is not the same as consideration of *wellbeing*, although it is frequently muddled with this, which can be considered under the No Harm Principle effectively.

It is a common mistake to think of beneficence as just another name for the No Harm Principle, or *vice versa*. This is not the case. They are two different concepts. The No Harm Principle requires us to recognise potential harm and devise ways to minimise the risks of it occurring; a simple way to put this is to say that the No Harm Principle simply means *do no harm*. Beneficence is directly concerned with *doing some good*. These are not necessarily the same thing. Just because you are doing no harm, you cannot say that this equates to doing good. If I avoid hitting you on the head with a hammer, am I really *doing good* to you? Yes, and no.

Mention has been made (in Chapter 5) of offering incentives or giving gifts to research participants. These are, of course, beneficial to participants or those associated with them. Within the social sciences there are no hard and fast rules in this area and you should consider taking advice from colleagues with more experience in the area when making decisions about how you may benefit others through your research. In terms of scientific research, materials are largely donated by individuals on a *pro bono* basis, with altruistic intentions (donating blood and other body materials, for example). Usually such donations are governed by legislation. Consider how any of these may be relevant or adapted to the dual use biosecurity context.

In the science context we need to consider beneficence in a slightly different way. The very nature of science with potential dual use means that work that is meant for the good of humanity, the environment, animal welfare and so on, may also be used to cause or facilitate severe harm to the same recipients. So 'good' or beneficial work has the potential, at the same time, to be 'bad' or harmful. Herein lies the dual use problem itself.

One obvious, if extreme, answer is to ban all science which has *actual* dual use applications, but the problem is that we do not always recognise or know what dual use applications may be possible either now or in the future. Another option is to ban all science which has *potential* dual use applications, but when and how will we know if it does have actual or potential dual use applications? A further option would be to control all scientific activity by some central monitoring agency that makes all decisions about potential or dual use applications unilaterally without the input of the scientists working on it.

Clearly none of these options is at all realistic or desirable. But this leaves us with the option of the life science community monitoring its own work for dual use risk and reporting centrally through some mechanism that enables the sharing of best practice, the promotion of risk identification and of mitigation. We cannot take the most superficially beneficial approach (banning all 'bad' science) because we will also lose the 'good' or beneficial effects of it. So we are left with a situation in which we need to work out how to maintain the benefits of life science research while minimising the potential for harm that may arise from the same science.

This is precisely why it is unlikely that a cost-benefit approach to weighing dual use science decisions is useful. There will *always* be beneficial aspects of a scientific process or activity that can be argued as more important than the need to avoid potential risks of harmful effects. Unfortunately, the cost-benefit claim is frequently the argument of choice when scientists are invited to (often publically) justify their work; what scientist is really going to 'cost' him or herself out of fame, fortune, prizes, increased funding, TV programmes and so on, in the interests of some unproven notion of risk to the public's safety? This is why we need independent, but *expert*, ethics panels composed of people *qualified* in ethics and in the relevant science, including informed lay people who represent the various views of the public on the wider concerns of society.

Being aware of this does not in itself help us to find a better approach, but we need to do so. This is not easy, but it is where we are, and where we will remain for as long as rogue elements among the human community want to misuse science for their own purposes. We also need to recognise that what one person views as rogue behaviour may be seen as entirely reasonable by another. How do we go about solving this dilemma? That is definitely for another book.

There is one other issue around beneficence that we need to look at. It is a thorny one. It is about us and the benefits that science brings to us personally and professionally. And it is about the way in which such benefits can cause us, sometimes, to assume a kind of paternalistic attitude towards non-scientists. This is an ethical issue because it can affect the way in which we look at other people and how we view their worries and opinions. Back in Chapter 1, I said that we can no longer rely on the argument that 'we know best' when confronted with the concerns of society about our work. Let us look at that a little more.

Do you think that all scientists are really disinterested (in both senses of the word) in the benefits that science can bring to them? Can you really say, hand on heart, that you are not pleased to receive the praise, prizes and plaudits that can accompany your scientific breakthroughs? Can you really say that you do not quietly enjoy being 'put on a pedestal' by your friends and family (plus by newspaper readers, journal readers, TV viewers, and radio listeners) when you produce some work that is recognised publically? To most of the non-scientist public, even having a single academic paper published is enough to put you 'up there' as if you are somehow more important than they. Many of my PhD students have expressed amazement at the way their status, in the minds of the public and of their friends and family, seems to change once they are known as doctoral students, and again when they become 'doctor'. This is just the way society is — but does it mean that we should get blasé about it and let such admiration start to affect our ethical judgement? Can you really say that this does not happen?.

Do you not occasionally look around at some well-known scientists and ask yourself 'What are they really in it for?' Like it or not, there is a trend of thought among the public that sees some scientists as being 'in it' for what they can get out of it, with little interest in the 'dangers' of their work. The fame and publicity that is afforded to some scientists, often due to the controversial nature of their work, plus what can come across as their arrogance sometimes, is often seen by the public as evidence that scientists seem to think that they are above the usual restrictions and concerns of society. I am not saying that this is actually true, but that it is a common interpretation among the public. We need to tackle this.

Fine, we all want to earn a living and we should be paid appropriately for our skills, knowledge and time. But do you not have a sneaking suspicion that sometimes fame and fortune can alter scientists' views on their own 'value' and values? How do we avoid arrogance and disdain for the 'uninformed' or 'stupid' concerns of the public? If we want to get the public onside, do not we have to show that we take their concerns seriously? Of course the public is not as 'informed' as we are about the technicalities of science. But does that make it right to dismiss their worries about what we may be doing as if they are children and we are wise parents?

What I am saying here is that we may need, at least occasionally, to think more about how our work can benefit the public — or not — and less of how our work can benefit us. This may mean that occasionally, rarely, we may be required, in the interests of the long term benefit of the public, to give up, delay or stop some of our work if it considered too dangerous or to have serious security implications. I would also say that we need to be big enough to let others — non-scientists — help us to make these decisions too. It is all too easy, when closely involved emotionally as well as professionally in our work to get to a point where all we can see is 'the work' (and what it can bring to us) rather than 'the bigger picture'. That's all I will say on this, but you get the idea. If you do not, maybe you need to read the last page again. Or at least, go away and think about it. If you disagree with what I have said, then at least have a coherent argument to explain why and do not simply rely on the 'scientific freedom' claim.

The Ethics Toolkit: Implementing Beneficence in Science

- Am I doing this work primarily for the benefit of others or for myself?
- How would I know this?
- How would others know it?
- What constitutes a benefit?
- To whom?
- What is the work for?
- Is the benefit real or just projected by me for some point in the future?
- How do we know that the benefit will ever be needed?
- Can we justify work now on the basis of a possible 'need' for that benefit in the future?
- Do I offer (wittingly or unwittingly) direct or indirect benefits to participants/research/work colleagues? (Apart from their pay and work benefits — see Chapter 5 — *Can I offer incentives to participate?*)
- Why am I doing this?
- Is this right?
- How may these 'benefits' be best managed (offered, limited, be appropriate)?

- Do they amount to coercion to participate or to continue with the work?
- Will the benefits actually lead to further problems to the recipients or to me?
- Are the benefits I offer colleagues actually meant to bring further benefit to me?
- Should I be offering certain benefits to colleagues in the interests of fairness?
- Have I been withholding benefits from colleagues or juniors in order to keep my place in the pecking order?
- Who resources benefits?
- Is there a conflict of interest? (There usually is if you look closely enough.)
- Where do the power relationships sit in this?
- Am I unduly influenced in my opinions on right and wrong in science by the benefits that my position as a scientist confers on me?
- How would I know that?
- Has anyone ever challenged me on this? How did I respond?
- If, from now on, all science publications had to be anonymous, and all prizes were abolished, what effect might that have on me and my attitude to my work?
- How would I manage my expectations about what science can do for me if I live in a country where scientists have to do what the government tells them to? (And cease from doing certain work if the government tells them to?)

Dual Use Biosecurity Ethics: Beneficence

Intervention Point 1:

You are working on benign research that may have dual use potential. You need to look out for this and take steps to minimise the risk of dual use being implemented by others. *How can the application of beneficence be a tool in supporting this aim? How may beneficence be a tool in enabling hostile dual use to occur?*

Intervention Point 2:

You do not know that you are working on projects that are being used for dual use purposes. *How can the application of beneficence be used to avoid this situation occurring and/or to protect the scientist and others? How may beneficence be a tool in enabling hostile dual use to occur?*

Intervention Point 3:

You discover later that you were, or still are, working on projects that are being used for dual use purposes. *How can the application of beneficence be used to avoid this or to mitigate the ill-effects of this knowledge causing harm to the individual?*

Intervention Point 4:

You are pressured to engage in dual use activities. *How can the application of beneficence be used to avoid this? What other employment or other social systems could be used to support workers, give advice and provide alternatives? Is any of this possible?*

If you are a scientist, how do you react to the Intervention Points *in terms of beneficence?*

What can you do, through the implementation of beneficence, to minimise the risks of:

- Being involved, or allowing others to be involved, in dual use work?
- Other people misusing your work?
- Other people engaging in their own dual use work?

Think about what level and type of beneficence, if any, needs to be obtained for:

- Yourself,
- Your colleagues,
- Associates of the work,
- End-users of the work/research,
- The public.

At which stage of research should your risk-minimising activities through beneficence be implemented:

- Research design stage?
- Data generation stage?
- Data analysis stage?
- Publication stage?
- Post-publication stage?
- Later? And where?

Who and what require protection from hostile use? What can you realistically do about it?

Deontology and Teleology of Beneficence in Science

Look at the Figure below and consider your responses to it.

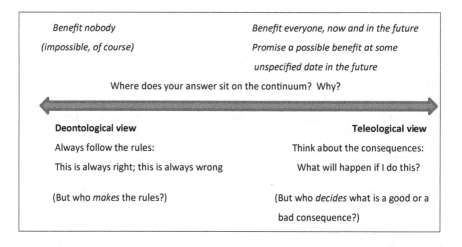

What considerations do you need to take into account when deciding where on the continuum your proposed answer lies?

The Responsible Dissemination Principle

In this context, I use the term 'dissemination' to refer to any activity that reveals your research, or any aspect of it, to those outside your team. This could include publishing, speaking at conferences, poster presentations, formal and informal discussions with colleagues, funding applications, allowing visitors to the laboratory, and any other way in which your work and its outcomes could become known. Of course, it also includes unauthorised access to your work. This is an ethical area that is often overlooked by scientists (and others). We assume it is our right and responsibility to publish our work, but in today's climate we need to think carefully, in some cases, how we may achieve this safely. It would be arguably unethical to restrict academic and scientific publication, but what about the potential harm publication may cause? This cuts straight to the heart of the scientific process, academic freedom and a range of other 'freedoms' and rights that scientists rely on as they act autonomously (see earlier sections).

There is an increasing awareness among researchers, especially scientists, that to publish their work as usual and open it up to scrutiny by society at large is to possibly open the way for misuse or misinterpretation of the work. This is particularly true in the dual-use context, where scientific papers may provide sufficient information to enable rogue individuals or groups (including nation states) to use the technology described for anti-social and hostile purposes.

Within the social sciences and health studies, published work may inadvertently cause social problems at a community level or wider. For example, some research may provide evidence that associates some minority communities with certain characteristics that may lead to stigma, to political fall-out or to civil unrest. While such research may be produced in order to enhance *support* offered to these communities, there are groups in society at large that would take the opportunity to use the research against them. I know, anecdotally, of at least one scientist with whom I have worked, whose health-based research has been refused publication on the grounds that it would not be conducive to the social good in the UK. I believe this work had been funded by the government or some major funding body. Whether you agree with this or not, the fact remains that some valid research may well produce results that cause, or could cause, problems socially, economically, or politically. I worked with another colleague who did some social research on the Royal Navy in the UK; this colleague had to sign the Official Secrets Act and could not publish the findings — which was the arrangement from the outset. I have other science colleagues who have deliberately not published at the earliest opportunity in order to secure some competitive advantage economically. Once they have secured a patent or some other economic control over the outcomes, they have published, even if not 'in full'.

We can see from these examples (and I am sure you will also know of similar cases) that 'not publishing' or 'partial publishing' is already a fairly common approach in science as well as in social science. The idea of restricted dissemination of findings is not new to dual use concerns. However, it must increasingly be borne in mind that our research could be wilfully misused for anti-social purposes. What responsibility do we have as scientists to mitigate the risks of this happening? Can we do this effectively? How do we self-monitor our responsibilities to society?

This issue goes to the heart of responsibility for the outcomes of scientific work. In reality, once work is in the public domain, it is beyond the control of the scientists who carried it out. In the light of the rapid advances now being made, especially in biological sciences such as neuroscience and synthetic biology (to name just two areas of rapid growth), it is increasingly incumbent on researchers to consider how, when and where they choose to disseminate their work. There are, unfortunately, groups of people aligned with political parties, nationalist or racist views, military or terrorist tendencies, and so on who will seek to access valid research to misuse it in the furtherance of their own agendas. In practice it is arguably impossible to maintain 'power' over your work once it is published. What would that mean to you on a personal level, as well as on a professional level if your work was ever misused in this way?

While we cannot be ultimately responsible for every application of our work, we need to consider how we may reduce the risk of misuse happening. The scientific community already protects its work through biosafety and other means, often due to commercial considerations. However, I would say that we now have to begin to consider these issues in more depth. Is it appropriate to publish your paper, or give a conference presentation, or a press interview, in *this* format, in *this* amount of detail, at *this* time, in *this* place, to *this* audience? How else may you publish or disseminate your work in a way that might reduce the risk of misuse?

There are no probably definitive answers to this, but spreading knowledge is a major ethical point to consider. Publication or other forms of dissemination impact directly on the application of the No Harm Principle and the principle of Beneficence. It may have an impact on the work of others as well as yourself. If you want to continue with your traditional view of scientific and academic freedom, how will you answer when that freedom leads to death and destruction? I know I am being challenging here, but somebody has to ask these difficult questions. Remember what we said earlier — it will only take one big event to reflect badly on 'our' science and *on we ourselves*. If we choose, now, to tackle this thorny issue 'in-house', we will be in the best position to maintain some useful level of control over our scientific activities in the future. If we choose to ignore this problem, or to dismiss it, then we will have an even bigger problem at some point in the future. Can we handle this? *How* will we handle it?

The Ethics Toolkit: Implementing Responsible Dissemination

- Am I aware of any current, recent or future potential for the misuse of my research?
- If so, how may I minimise the risk of such misuse through consideration of my dissemination methods?

- Are there options to be explored about how I could disseminate my work in other ways than the traditional routes at this time?
- Who can advise me on alternative dissemination methods?
- If I choose to redact or partially publish, what will I *really* suffer?
- What *advantage* may I gain through only partial publication?
- What other arrangements could I put in place to supply details to other scientists after I partially publish my work?
- What advantages could this give me?
- What disadvantages could it give me?
- Who will potentially be affected by any dual use of my work? Should they be warned?
- What impact would full or partial publication of my work have at a different time or place?
- Should I consider holding back some of my findings from certain types of publication or from certain groups?
- How would I disseminate those withheld findings to 'safe' destinations?
- Who decides what is a 'safe' destination?

Dual Use Biosecurity Ethics: Responsible Dissemination of Research

Intervention Point 1:

You are working on benign research that may have dual use potential. You need to look out for this and take steps to minimise the risk of dual use being implemented by others. *How can the responsible dissemination of research be a tool in supporting this aim? How may responsible dissemination considerations be a tool in enabling hostile dual use to occur?*

Intervention Point 2:

You do not know that you are working on projects that are being used for dual use purposes. *How can the responsible dissemination of research be used to avoid this situation occurring and/or to protect the scientist and others? How may the responsible dissemination of research be a tool in enabling hostile dual use to occur?*

Intervention Point 3:

You discover later that you were working on projects that are being used for dual use purposes. *How can the responsible dissemination of research be used to avoid this or to mitigate the ill-effects of this knowledge causing harm to the individual?*

Intervention Point 4:

You are pressured to engage in dual use activities. *How can the responsible dissemination of research be used to avoid this? What other employment or other social systems could be used to support workers, give advice and provide alternatives? Is any of this possible?*

If you are a scientist, how do you react to the Intervention Points in terms of the responsible dissemination of research?

What can you do through the responsible dissemination of research, to minimise the risks of:

- Being involved, or allowing others to be involved, in dual use work?
- Other people misusing your work?
- Other people engaging in their own dual use work?

Think about what level and type of responsible dissemination of your work, if any, needs to be obtained for:

- Yourself,
- Your colleagues,
- Associates of the work,
- End-users of the work/research,
- The public.

At which stage of research should your risk-minimising activities through responsible dissemination of your work be implemented:

- Research design stage?
- Data generation stage?
- Data analysis stage?

- Publication stage?
- Post-publication stage?
- Later? And where?

Who and what require protection from hostile use? What can you realistically do about it?

Deontology and Teleology of Responsible Dissemination of Research

Look at the Figure below and consider your responses to it.

Where do your 'beneficence' answers to these questions sit on the continuum? Why?

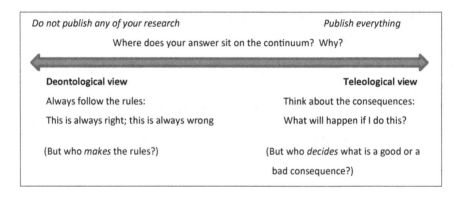

What considerations do you need to take into account when deciding where on the continuum your proposed answer lies?

The Scholarship and Continuing Professional Development (CPD) Principle

This is arguably (in some peoples' view) not an ethical issue in itself, but one of academic or professional practice. There is, however, an argument that it is *unethical* to carry out and produce work that is less than competent, well-informed, honest, reliable and valid. This presupposes an appropriate level of qualification, experience and competence in you as the scientist. This makes it, in my view, an ethical issue. As scholarship skills and research methods training are covered by training in academic practice and in research methodology training (where it exists), it is not proposed that they should be covered in detail here. But we do need to think about the ethical implications for us as scientists because the public assumes that we *are* totally competent in what we are doing, and that includes being ethical from an informed and up to date stand point.

It is therefore an ethical responsibility of every researcher and working scientist to maintain his/her skills base, to seek to always enhance his/her knowledge of the relevant fields, to build on his/her methodological expertise and to practise in an honest, transparent and open way with research subjects, their associates and colleagues. If you agree with this, can you make a list of all the activities that you have involved yourself in over the last year (for example) which have been specifically undertaken to update you and to advance your skills and knowledge? What about the same sort of activities offered by you to your staff and juniors in order to pass on best practice and new ideas?

To engage in scientific practice is to engage in socially responsible activities. The maintenance and increase of your skills and knowledge is an ethical requirement if you are to continue to practice effectively in your chosen field. Yes, we learn a lot 'on the job' once we have gained our qualifications, but we also need to build up our training as we progress professionally. We need to be brought up to date with new advances. We need to hone and develop our skills and acquire new ones. This applies to ethical competency as well as to competency and expertise in our chosen field. Plus, we need to pass this on to the next generation and to our peer groups.

If we are employed to carry out work, our competency is implied. If we are funded to carry out work, our competency is implied. Our work affects people, animals or the environment. It may affect all three groups. Can we afford to be less than competent and less than expert in what we are doing? Our expertise and competence forms part of our defence if and when things go wrong. We ought to nourish and protect our skills, knowledge and experience in order to be better practitioners. To be in a position to hand our work on to the next generation in as ethically-sound a form as possible is surely one of our greatest responsibilities. Ethics training needs, therefore, to be a part of the education of every science under-graduate, postgraduate, post-doctoral scientist and professional scientist. This can be facilitated by effective, level-specific ethics education through CPD sessions at regular intervals in your facility. Opportunities to share best practice between facilities, for example through meetings of professional associations, are also vital as tools to improve standards and to achieve 'buy-in' to ethics from scientists at all stages of their careers. Can we really afford *not* to do this in today's world?

The Ethics Toolkit: Implementing Scholarship and CPD

- How can I keep up to date with methodological developments in my field?
- What further training and development do I need to work more effectively?
- How will I know that I need to be brought up to date?
- How much of my time should I spend on CPD?
- Do I have any responsibility to maintain the CPD of my colleagues/staff?
- How can I make applied ethics a regular part of the learning of my juniors and colleagues through CPD sessions?
- How can I spread my/our good practice (ethics and science) to other colleagues?
- How can I/we pick up the good practice (ethics and science) of colleagues and bring it into our facility?

- How open am I to change my practice (ethics and science) if I see a better way?
- If I am introduced to a more ethical way to carry out my work, am I prepared to change my work accordingly?
- Am I prepared to talk to my colleagues to persuade them to change their work?
- What issues are preventing me from changing my practice?
- What issues are preventing my colleagues from changing their practice?
- What can we do about this?

Dual Use Biosecurity Ethics: Scholarship Issues and CPD

Intervention Point 1:

You are working on benign research that may have dual use potential. You need to look out for this and take steps to minimise the risk of dual use being implemented by others. *How can up to date scholarship skills and CPD be a tool in supporting this aim? How may up to date scholarship skills and CPD considerations be a tool in enabling hostile dual use to occur?*

Intervention Point 2:

You do not know that you are working on projects that are being used for dual use purposes. *How can up to date scholarship skills and CPD be used to avoid this situation occurring and/or to protect the scientist and others? How may up to date scholarship skills and CPD be a tool in enabling hostile dual use to occur?*

Intervention Point 3:

You discover later that you are/were working on projects that are being used for dual use purposes. *How can up to date scholarship skills and CPD be used to avoid this or to mitigate the ill-effects of this knowledge causing harm to the individual, the science profession and the public?*

Intervention Point 4:

You are pressured to engage in dual use activities. *How can up to date scholarship skills and CPD be used to avoid this? What other employment or other social systems could be used to support workers, give advice and provide alternatives? Is any of this possible?*

If you are a scientist, how do you react to the Intervention Points *in terms of up to date scholarship skills and CPD?*

What can you do *through the promotion of up to date scholarship skills and CPD*, to minimise the risks of:

- Being involved, or allowing others to be involved, in dual use work?
- Other people misusing your work?
- Other people engaging in their own dual use work?

Think about what level and type of up to date scholarship skills and CPD need to be obtained for:

- Yourself,
- Your colleagues,
- Associates of the work,
- End-users of the work/research,
- The public.

At which stage of research should your risk-minimising activities through the maintenance of up to date scholarship skills and CPD of your work be implemented:

- Research design stage?
- Data generation stage?
- Data analysis stage?
- Publication stage?

- Post-publication stage?
- Later? And where?

Who and what require protection from hostile use? What can you realistically do about it?

Deontology and Teleology of Scholarship Issues and CPD

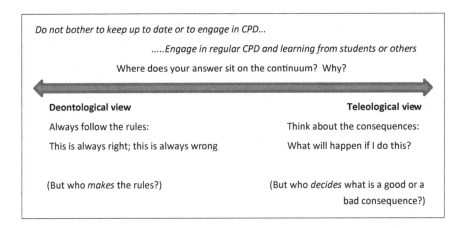

Do not bother to keep up to date or to engage in CPD...

.....Engage in regular CPD and learning from students or others

Where does your answer sit on the continuum? Why?

Deontological view   **Teleological view**

Always follow the rules: Think about the consequences:

This is always right; this is always wrong   What will happen if I do this?

(But who *makes* the rules?)   (But who *decides* what is a good or a
                                              bad consequence?)

Where do your 'up to date scholarship skills and CPD' answers to these questions sit on the continuum? Why?

What considerations do you need to take into account when deciding where on the continuum your proposed answer lies?

Here, we have reached the end of The Ethics Toolkit questions. In the next chapter we will look at when The Toolkit should be applied, and when the ethical status of your work should be reviewed.

Chapter 7

How and When to Apply
and Review Ethics in Practice

Having looked at the ethical principles in some detail in the previous two chapters, this chapter looks at *how* and *when* to apply ethics in a broader sense. It is all very well gaining an understanding of the principles themselves, but we also need to look at the bigger picture in terms of making the application of ethics a significant part of our planning and review policies and activities. In this chapter, I will introduce the various gateways 'into' ethics that you can choose from as you plan and review your work (and that of others). We will also consider how and when to carry out Ethics Reviews.

The Four Gateways to Identifying and Addressing Ethical Issues

There are at least four ways of approaching the ethical principles (and in this case, recognition of dual use potential) as a scientist:

- By looking at each ethical principle in turn and applying them to all people and situations concerned at each stage of the research process;

- By looking at the stages of the research process first and then considering what ethical principles may require accommodation at each stage;
- By starting with an ethical problem that you know of and then seeing how it may affect/is affecting your research and even the research design, from the beginning then making changes to your work;
- By taking a relationship approach.

These approaches may be viewed as four different gateways into working with the ethical aspects of your research. Whichever gateway you choose (and you may use more than one), it is very important that you work your way through the entire research process to look ahead (or backwards) to 'see' what ethical principles may be compromised at any or all stages of your work. In this way you are most likely to be able to recognise and deal with any difficulties before it is too late to make changes or avoid ethical problems.

Ideally, it is best to consider ethics *from the very beginning* of your research plans. It may be that an ethical issue is so insurmountable that you have to change your research design to accommodate or avoid the problem. It is much easier to do this at the beginning or at an early stage than to have to juggle all your plans and arrangements (which may have funding implications) later. By then, you may have caused significant harm, even unintentionally. This could include professional harm caused to your reputation as a scientist.

Of course, it is always possible that even with the best planning in the world, some unexpected ethical dilemma may emerge during the research process. My suggestion, should this arise, is that you *do something about it* immediately, as soon as you recognise a potential ethical problem. Do not ignore it and hope that it will go away. It will not. Ignored, it may ultimately have a hugely negative impact on the research, the people affected now and in the future, your future work and future funding opportunities and even the future work and funding of others. In terms of dual use, to ignore some potential once it has been identified has even more serious consequences. As soon as you become aware of any potential for the harmful dual use of your work, then it is your responsibility to take some action to mitigate the risk of it being carried out. Ideally, you should flag up the possibility of misuse with your team and with your bosses as soon as you spot it. This could turn out to be *the* vital defence policy if you are ever challenged in the future. You should, of course, also inform the relevant Ethics Committee/Panel. You should also inform your funders and perhaps your professional association. All of these have an interest in preventing misuse of your work.

By taking an ethics approach to your work at all stages, not only will you have an extra resource to help in identifying potential problems (The Ethics Toolkit), you will also have a road-map that will guide you in deciding on what to do next. The Toolkit will not provide you with specific answers in detail, but it will:

- help you to identify potential or actual ethical problems,
- help you to identify who may be, or is, affected,
- help you to identify where, or with whom, you could take mitigating action to reduce risk,
- help you to check, after taking mitigating action, if your action has been appropriate and effective (by reviewing the ethics and outcomes with the mitigating action in place).

Who should be Involved in Ethics Reviews and Planning?

In practice, it is unlikely that you will have to make such decisions alone in any case. It is always advisable to 'plan' your ethics approach to your proposed project with the help of someone who is qualified in some way in applied ethics. Speaking to someone who has carried out similar research successfully may help (but are you sure they acted ethically?).

I would advise anyone who is in a position to devise research to take some classes or some sort of tuition in applied ethics as soon as possible. Not only is it important that you get a grip of applied ethics for your own study, you also need to be able to pass on this knowledge and skill set to others in order to build ethics capacity in the life sciences. In my experience, nothing matches the level of learning that we experience when we see some aspect of ethics *in action* in the real world.

You will remember the principles and the solutions much more easily when you see them being applied and used, because you associate them with a real life experience. At least reading this book will start you off in the right direction — you can use The Toolkit questions as a good guide in the absence of any other training. But I would still advise that you try to find some applied ethics courses that focus on your area of research.

Why not engage with a member or members of the relevant ethics board at the planning stage (or any other stage) of your project? In this way, the board will be already aware of your proposals, and by seeking help early, you are likely to avoid delays and problems later when you submit your project for approval. As I said earlier, if the board know that you are actively engaging with ethics from an early stage, they will be more inclined to help you and to support you. You would not usually consider re-wiring your house without the help of a qualified electrician — so why not seek the help of an ethics expert at all the important stages of your project?

Try to do the 'gateway' approach with your team from the beginning, or at least from the earliest stage in which you have partners in the project. This will ensure that everyone is 'bought in' to the ethical aspects of the project from the day they join the team.

Gateway 1: Identifying ethical issues first
Review the ethical issues as described in previous chapters:

- Autonomy (self-determination) — the rights of individuals
  o Voluntary participation,
  o Consent,
  o Privacy — anonymity, confidentiality, time, and place.

- Your direct responsibilities as a scientist
  o Doing no harm,
  o Doing good — beneficence,
  o Responsible dissemination of your work and findings,
  o Scholarship issues and continuing professional development (CPD).

Recognise how these may impact on what you do, where, when, and how you may carry out your project. Some ethical dilemmas may mean that you cannot undertake the project you wish to, as the ethical issues are insurmountable.

Then, think of the stages of your research process. Map the ethical principles onto the stages of a research project (as described in Gateway 2). Consider which ethical issues may impact on each of these stages of the research process. How may the ethical issues you recognise actually *form* your research plans?

Devise your research plans and the management of the research stages accordingly. This should be an ongoing practice throughout the research process.

Gateway 2: Devising the research design/stages first

Look at the various stages of your proposed research project:

- Formulating the research question(s),
- Deciding on a methodological approach,
- Designing the study,
- Getting funding,
- Recruiting the full team,
- Sampling,
- Generating data,
- Reviewing and analysing data,
- Reviewing findings,
- Further stage iterations,
- Publication,
- Post-publication responsibilities.

Once the research stages are planned in outline, go back to the ethical issues above (Gateway 1) and see how they (all or some) may be mapped onto *each* of the stages of the research. You may find that you have certain ethical issues at only some stages of the research. You may also find that you have one or two serious ethical issues that require you to redesign parts of your project to minimise the risk of harm — especially of bad outcomes. You can use this 'mapping' technique to help you to reduce the risks of misuse of your work in the future.

Make changes to the research plans accordingly. This should be an ongoing practice throughout the research process.

Gateway 3: Starting with a knowledge of a specific ethical issue

There may be occasions when you know of certain ethical dilemmas, or potential dilemmas, *before* you begin planning your research.

When this is the case, you can use elements of the first two gateways in devising a way forward that accommodates the needs of both yourself and your research participants/colleagues, plus the safety of the wider public.

- Identify the ethical issue(s) you know of or suspect may be about to arise (from Gateway 1),
- Consider which stages of research the issue(s) will or may map onto (from Gateway 2),
- In what way do the recognised ethical issue(s) impact on your proposed research plans? Is the impact felt at every stage or just at one stage?
- Can the ethical issue(s) be reasonably minimised or avoided?
- If not, what can be done in terms of the research planning and undertaking to accommodate it/them?
- What stages of the research project will be affected and need to be modified?
- Does the existing ethical issue mean that your research question itself is untenable?
- You may have to make use of an alternative methodology or data collection method(s) — can you do this and still achieve the desired aims of your project?
- Be prepared to radically change your plans if the ethical issue(s) cannot be accommodated in your plans; you may be able to return to the project at a later date when some other approach is possible,
- You may have to abandon the research if the ethical issues are insurmountable,
- Once you have addressed the original ethical issue that you knew about in advance, take Gateway 1 or 2 and review the rest of the work plans for the other ethical issues,
- You may have to consider restrictions in publication of your work,
- If this arises, devise effective ways to allow access to your work by those who have a valid need to see it — this will most likely require external advice and support.

Ultimately, you may have to abandon or postpone your planned research, or at least make significant changes to your initial plans. But it is better to do this than to be exposed to censure later when your work has been misused and you are shown to have failed in your ethical precautions.

Gateway 4: Taking a relationship approach

Sometimes it is useful to simply consider ethical issues and your research plans from a relationship perspective. This can be a quick way to identify where ethical issues may arise, and what those issues may be. You can often use this gateway as a 'check up' method to review your planning when you have used one of the other gateways.

- Researcher — participant: how might my research breach his rights or cause him harm?
- participant — researcher: how might something about him adversely affect my research?
- Researcher — researcher: how might my work adversely affect the current and future work of other researchers?
- Researcher — reader or public audience: how might my research affect the public — beneficially or adversely?
- Researcher — family (own and participants'): what effects does my work have on my family and their families?
- Researcher — colleagues (own and participants'): what effects does my work have on my colleagues and those of my participants?
- Researcher — wider society: what effects does or may my work have on the world at a population level?
- Finally, having identified potential issues this way and addressed them, use Gateways 1 or 2, or a combination, to review all of the stages of your research against the ethical principles.

By looking at the links between you as a researcher and others, ethical issues naturally begin to emerge. This enables you to consider how the issues may impact on the various stages of the research process. This will, in turn, allow you to work on ways to accommodate them. As ethics is all about how we relate to others, this can become a favoured way of approaching the issues effectively.

Ethics Approval Processes 'Get in the Way'

Now that we have looked at the ethical principles themselves, plus the various gateways through which you can proceed to deal with them, let us consider *when* to apply them. It is all very well saying that we have 'done' the ethics, but ethical behaviour does not actually work like that. We need to keep on 'doing' ethics all the way through our work and beyond. This means reviewing 'the ethics' — or rather, the *ethical status* of our work — both at regular intervals and when unexpected changes arise.

At the time of writing, I have been teaching research ethics in higher education for nearly 15 years. Most of my students have been PhD candidates, a few have been Masters students. Some have been staff researchers. They have come from a range of disciplines within both the social sciences and the natural sciences. The most common misconception that I have met among these groups has been what I call the 'tickbox' approach to ethics. In this, the student thinks 'Once I have got ethics approval for my project I can really start the work'. Usually this has been echoed by the PhD supervisor or the PI. This says, in effect, that once ethics is 'ticked off' the list, then work can really start. The prospect of returning to the ethics is never considered because 'it has been done'.

I know of one institution where an ethics approval policy was instituted after years of argument — mainly between scientists and social scientists who needed different support from the policy and had conflicting ideas of what 'ethics' means. This new policy required research supervisors (and staff PIs about to carry out research projects) to look first at the 'Ethics Inventory'. The idea was that if the Ethics Inventory was signed off by the supervisor or staff PI as being non-contentious, then full ethics approval did not have to be sought. All signed Inventories had to be sent in to the central research office; here, they would be reviewed by the Chair of the relevant approval panel, signed off (or returned with a question) and filed. Cases requiring full approval would be referred to the full approval process. I never got to the bottom of what framework or process was used to make these initial decisions.

The 'Ethics Inventory,' followed by recommendation for some projects to 'go' for full approval, sounds good in theory but is still full of pitfalls. First among these is the qualifications, knowledge and experience of those in authority in the ethics approval process. The Inventory I mentioned above was heavily dependent on the ethical competency of the supervisor or the staff PI in the first place. Further, in my experience, a structured ethical expertise is often lacking among the members of some ethics approval panels (I am talking about many institutions here), let alone among PhD supervisors and PIs. What happens in practice tends to be the application of the personal values of the members of the panel, or the supervisor, to the proposed research or work. This is *not* the same as the application of a structured ethical interrogation of a proposal. I am not saying that the experience of supervisors or panel members is worth nothing, but simply pointing out that not one of those colleagues with whom I have discussed this over the years has been able to point me to the ethical framework or structure against which he or she assesses proposals. Experience can only go so far.

Problems with policies and power relationships

At the institution I mentioned above, this new ethics policy was hailed as a breakthrough in the institution's ethical responsiveness. In so far as it *was* an ethics policy (where there had not been one, or at least an effective one, previously), I would agree with this view. However, just because a policy is in place, we can't assume that it is a good and effective policy, or that it solves all ethical problems. Neither can we assume that everyone will engage with it properly. A policy is only as good as the way in which it is implemented; even this assumes that it is a good policy in the first place. In this case, as there were no effective sanctions in place with which to counter any non-engagement or failure to comply with the requirements of approval, the policy had 'no teeth'. People could, and did, ignore it or manipulate it. When I and others proposed a blanket ban on the progression of projects until full ethics approval was given (amongst those projects that were deemed to require full approval) — which would only have taken a maximum of six weeks — we were overruled in the face of a revolt amongst staff. What does this say about the value and understanding of research ethics in a research institution? In effect, we were expected to allow, indirectly, the progression of projects which had already been deemed potentially ethically problematic, because ethics was seen as 'getting in the way'. In a three or four year project, what is six weeks if planned into the schedule? This attitude displayed a failure to appreciate not only the risks to the ethical quality of the work, but also professional risks to those undertaking it and to the institution itself. Considering that the institution would need to fund any compensation claims or lawsuits arising from unethical research, this is astonishing. But common, unfortunately.

I regularly had students from my last university coming to me saying that their supervisors had told them to 'just do the tickbox and get the ethics done' without actually looking at any possible ethical problems that may arise in their projects. This is easy to explain: the supervisors had never had ethics training. Some may have attended a session once in the dim and distant past, which perhaps focused on a very general approach to ethics, but such sessions tend to be of little use in my opinion, because they remain largely theoretical. Sessions looking at specific issues such as the ethics of x or y (often public awareness-type sessions) may help occasionally in terms of raising awareness of x or y, but usually fail to show *application techniques*. I know I am generalising here, but this is my experience. Most people will be interested in the theory of ethics in society, but then get stuck at how to apply it to their own work.

The PhD students (including some staff researchers) who were required to attend my classes had more ethics training than did their seniors. Here we meet our old friend again, the power relationship. How is a student or a contract researcher supposed to question the uninformed decision of his or her supervisor or PI? More to the point, when the junior actually recognises a potential or actual ethical problem in the project, to whom can he or she go for advice and support if the supervisor or PI does not accept that judgement? I had a steady stream, for years, of 'old' students knocking on my door long after they had completed my courses, stuck for somewhere to go for help. Some went to the members of the panel that had approved their project, others went to other staff in their faculties. Many ended up in the chair on the other side of my desk. This occurred not because their seniors were unhelpful *per se*, but because their seniors did not have the knowledge or skill set with which to advise and help them. As I said in an earlier chapter, ethics is probably the only subject that everyone believes they already have a full knowledge of without being taught — because they confuse ethics with their private personal values.

The problem with the Inventory in question was that most supervisors were of the opinion that ethics is something that 'gets in the way' and is just a hold-up *en route* to 'getting on with the real work.' The other issue, related to power and academic status, involved supervisors not wanting to appear to be uninformed about ethics in front of their students. It was therefore the aim of many supervisors to fill in the Inventory in such a way as to make the research sound as innocuous as possible in order to avoid having to apply for full ethics approval. I know this because students repeatedly told me; further, some supervisors (from a range of disciplines) used to periodically come to my office to complain about the nuisance of the ethics approval process.

The thought that ethics may actually be a normal and full *element* of research seemed to be missed. Neither was it ever recognised as a means of professional defence should serious problems arise. To add to this view, the misconception of ethics as a one-off tick box activity at the beginning of a project (never to be thought of again) simply resulted in the maintenance of an anti-ethics perspective among supervisors and many PIs. This was problematic because many students and researchers picked up this view and had their perspectives on ethics diluted accordingly. This is what I refer to as 'ethical erosion' and is to be regretted, as it can only expose students and qualified staff to accusations of ethical incompetence, or worse, if something later goes wrong. This negative view of ethics appears to be a common view everywhere I have worked, both in the UK and overseas. I am sure you recognise all of it.

An Exemplar Ethics Inventory — With Problems

Let us look for a moment at an exemplar Ethics Inventory that *looks* good but actually contains some unrecognised pitfalls. All of these questions are taken from several real ethical approval forms that I have

---

Is the proposed project an empirical research project involving people?

- Will the project include primary data collection from human participants, their data or their tissue?
- Will it constitute an 'investigation undertaken in order to gain knowledge and understanding'? (This includes work of educational value designed to improve understanding of the research process.)

CRITIQUE

Define 'empirical'. This may sound crazy, but I have seen all sorts of arguments to avoid this designation. In any case, what has the 'empirical' nature of research got to do with the rights of human participants? The participants require protection whether research is empirical or theoretical. Even an overview of multiple projects such as in meta-research can cause ethical problems for the people who are associated with the topic if its findings may be contentious. The question should simply relate to investigation of people, their views, beliefs, characteristics and so on, rather than the *nature of the research* itself. Both approaches (empirical and theoretical) can produce 'difficult' findings.

Define primary data. Even after training, many students (and seniors) still get confused about the difference between primary and secondary data. If you don't believe me, do a survey among your students or colleagues. This always needs checking.

And finally, isn't *all* inquiry undertaken to 'gain knowledge and understanding'? Why do it otherwise?

If you answer 'Yes' to the first question, ethical approval <u>may</u> be required.

CRITIQUE

Ok ..... (the form-filler is now starting to sweat)..... he is thinking 'How can I exploit the 'may' in this?'

If you answer 'No' to the first question, then a research ethics review is not usually required.
Note: there may be occasions where a project is not defined as research but still raises ethical issues — please submit for review if this is the case.

(Continued)

(Continued)

CRITIQUE

Hmm. The problem here is that this question and note fails to take into account the possibility that even new research using secondary data could do real harm. Secondly it assumes some inquiry (define 'research'!) is likely to be 'ok' and will pose no ethical problems (the comment is loaded that way). Thirdly, the form-filler is now being asked to identify ethical issues himself — which he may not be able to do. Overall, this sort of comment suggests that 'non-research' inquiry, such as an audit, does not require ethical scrutiny. But it may need it.

Is the proposed project an audit or service evaluation involving humans?

CRITIQUE

Define 'audit' and 'evaluation'. In my experience, neither is without ethical problems. There seems to be an assumption in many areas that audits do not require ethical approval. I am not saying that they all *do* require full ethical approval, but this attitude fails to recognise that an audit, not to say an evaluation, can raise a number of ethical issues itself. Who is doing the audit or evaluation? What is their competency? What are their qualifications? What is their level of understanding of the issues faced by those being audited or evaluated in their work? Are the right questions being asked and the right observations being sought and included? Is the audit or evaluation being undertaken at an appropriate time? Is the methodology of the audit or evaluation an appropriate choice? What possible outcomes may arise for those being scrutinised? These are all ethical questions and can all be related to the principles outlined in previous chapters.

Will the research project involve the National Health Service?
If you answer 'Yes' to this question, ethical approval will be required by NHS Research Ethics Committee (NREC).

CRITIQUE

This is an example of the requirements for many researchers/scientists to meet the demands of external governance bodies. But does approval by an external body guarantee that all the ethical issues have been recognised and addressed? (Hint — no). See below for further comment on this.

If you consider that full ethical approval is not required, please explain briefly why not.

(Continued)

(Continued)

CRITIQUE

This assumes that the form-filler is ethically competent, when he may not be so. The answers may *look* ethically sound, but how does the reviewer know that the writer has not just used all the 'right' words? The question can also raise unrealistic expectations — some form-fillers will assume that because they have some 'valid' reasons (in their own eyes) to avoid ethical approval, they will subsequently be able to do so. When questions then come back to them, it simply raises the level of resentment and ill feeling towards 'ethics'.

Will the research project involve any of the following:

- **Testing a medicinal product**
- **Investigating a medical device**

CRITIQUE

Such activities are usually governed by external bodies, regulations and law, but still require specific ethical scrutiny. Don't rely simply on 'outside' approval and think that your internal scrutiny is therefore covered. It may not be.

Will the research project involve any of the following:

- **Taking samples of human biological material (e.g. blood, tissue, hair, cell lines)**
- **Prisoners or others in custodial care (e.g. young offenders) as participants**
- **Adults with mental incapacity as participants**
- **Children under 18 (the legal age of majority in the UK)**
- **Other vulnerable groups as participants**

CRITIQUE

This is all about consent and the level of information provided to consenters. All of these classifications of participants require the application of all of the principles outlined in previous chapters because of the autonomy of the individual. In the case of minors or of adults with limited mental capacity, third party consent is required. This is not always straightforward. Third party consenters carry a heavy responsibility. Even samples obtained from deceased persons will have had to go through some form of consent before they get to you at your bench. Are you satisfied that this has been done properly and to a high standard? How do you know that third party consenters are really working in the best interests of the individual? There will also be several legal hurdles to get over in order to work with these groups as well. Do any of your ethical responses clash with your legal responsibilities? If so, which will win out?

Will the research project involve animal testing or other use of animals?

(Continued)

(Continued)

CRITIQUE

This is not the place to get into the debate of ethics and animal testing. Suffice to say, it is an ethical minefield. Why? Lack of CONSENT and respect for all life! Then we have the rights of animals to be treated well; to be fed and watered; to be warm and safe; to have a full life and so on. However, having read this far, you are likely to be aware of the competing priorities of rights that must be balanced in any argument between the rights of humans and the rights of animals. I don't like animal testing, but I recognise that it may have to go on until we can figure out effective ways to do without it. That does not mean that we should not require the highest levels of ethical behaviour for animal welfare in such cases.

Will the research project involve human participants and/or human data that is acquired through routes that do not fall under other authorities (such as the NHS)?

1. Through interviews, questionnaires, surveys, observations, etc.?
2. How many interactions will take place with each individual?
3. How long will the interaction last?
4. Who are the participants?
5. Will consent be sought?
6. Where will interactions take place?
7. Attach all documentation including any proposal, consent forms, information sheets, interview guidelines, questionnaires/surveys, etc.

If you answer 'Yes' to this question, then full ethical approval is required. Attach all documentation including any proposal, consent forms, information

CRITIQUE

A lot of this is very 'social science', but some forms of data, such as 'observations' can cover a multitude of activities, even in a lab. Just because you are observing something in a lab, relatively far removed from the actual person who donated the sample, you are not exempt from looking at issues of consent, privacy, no harm, beneficence and so on. Consent must have been obtained somewhere at some point. Question 5 here (consent) is typical on many forms, unfortunately — it assumes that consent may not always be necessary. If we are dealing with humans, consent somewhere is necessary. It may be required to be re-gained at various points during the research. How would you achieve this? Question 2 (how often will the research/er interact with the individual), is a good one. It concentrates the mind of the form-filler to expect further questions.

I have discussed this project with my student/colleagues

(Continued)

(Continued)

CRITIQUE

Right....and? Your discussions may have consisted of 'this is all a load of rubbish but we have to fill it in — anyone know any fluffy words?' Effective discussion should require the signatures of all of those consulted, to go *some* way to prove that discussion and agreement have taken place. Unfortunately, power relationships appear again here. 'Discussed' may actually mean 'I have told my staff to sign this form' and in fact, no discussion or debate has occurred. Worse, the team disagrees with the PI but the members are still pressured by the PI to sign this off. In either case, signatures are meaningless, but *look* as if they mean that all concerned are satisfied with the form and the project. Juniors beware.

I confirm that there are no ethical issues requiring further consideration

CRITIQUE

This is often added to forms but should be avoided. Any form allowing this box to be 'ticked' is a hostage to fortune; how do you know that the form-filler is ethically competent? How do you know that he has made the right decision? How do you know that he is being honest? What about unforeseen ethical problems that will arise during the research? What adverse outcomes may arise from the wrong 'ticking' of this box?

I confirm that there are some ethical issues that require further consideration. I will be requesting full ethical approval.

CRITIQUE

Do you know anyone who would willingly tick this box? Hats off to anyone who does.

Does the project have the potential to produce results or findings that may be misused for hostile purposes against humans, animals or the environment?

CRITIQUE

This is one for the future — I have never seen it on a form yet. If only this question were to be included on all ethics approval forms, we would be some way towards accepting dual use as a real issue. Of course, it presupposes that dual use is actually looked out for — which requires training.

seen in the UK. I have added my own critique of each section throughout the 'form'.

What do you think now that you have read this?

As with all assessments made by form-filling, the work is being judged here largely on *how well the form has been filled in*, rather than the quality of the work under review. We also need to be aware of people who seem to know all the fluffy words (we have all played 'lingo bingo' in class and when marking assignments) and include them in the form, but who do not really understand what they mean in practice. It is always useful to define your terms in paperwork so that all involved have to work to the same definitions.

I will not go into further detail here about the full ethics approval process as you should be able to devise your own having read the previous chapters and added the content to your own experience. Suffice to say that I believe that we are under-checking research and that many institutions with Ethics Inventories (or equivalent preliminary stages in the approval process) will probably be unwittingly letting projects through that ought to have been subject to more detailed scrutiny. This is potentially dangerous, for individuals and for institutions. As I have said earlier, we only need one bad case in which poor ethical standards result in a public incident, followed by an institution being fined or otherwise prosecuted, for minds to be concentrated on this.

Reviewing the Ethics of Your Project Post-Approval

The last point to make about the process itself is that we also need to build-in regular reviews of the ethics during each project. All approval processes should include in them some recognition of the need for this, preferably with a list of times and circumstances when review should be carried out. This will occasionally require a further ethics approval process — but remember that the long term aim of such processes is the protection not only of your project but of *you*. If we build-in time for ethics approval, this does not need to interfere with schedules.

The thing to remember is that we are dealing with *people* and people can change their minds or their circumstances. Consent is not always a one-off issue. It may need to be revisited often. All aspects of autonomy require review if research interventions go on for more than a one-off sample donation or other element of inquiry. People who engage with us as researchers have the right to withdraw their consent at any time. They can also withdraw their data or whatever other material or intellectual property they have donated to us. This all requires consideration. I know that as a lab-based scientist some of this may sound unfamiliar if not excessive, but as we have considered earlier, science cannot continue in its own bubble in the same way that it has arguably done to date.

We need to get used to the fact that our staff, our students, our colleagues and the public all have an expectation that our work will only be used for good. We cannot go on assuming that nobody will misuse our work, unfortunately. So we have to take steps to minimise this risk. Ethics is a good way to start.

Gold Standard Ethics Approval — A Cautionary Comment

I mentioned above (in the Inventory critiques) some issues with external bodies and ethics approval. I will talk about the UK's National Health Service here, but you could apply this to the influence of any national health care body anywhere (or indeed to the ethical judgement of any professional association or body that is held up as a gold standard). Take a look at Figure 7.1 and see what you think. In the UK, NHS ethics boards approve or reject research activities on behalf of *patients* and *future patients*. They are concerned with guarding the rights of, and benefits for, people who are ill, or who are going to be ill in the future. This means that there is a cut-and-dried rationale for much research to go ahead: potentially we may all benefit from it because we all get sick.

I have noticed in the UK that there is often a tendency to think that if something has been approved by an NHS ethics board, then this approval is acceptable *and sufficient* in other research contexts. I don't always agree with this. I am not saying that NHS ethics approval is worthless, far from it. What I am proposing is that ethics approval in a health care setting may be arguably a different entity to ethics approval in another setting. Take a look at Figure 7.1 and see where possible differences lie. It may be that certain actions are prohibited in an NHS setting that may be ethically allowable in other

Health care research	Scientific research
Assumes as a foundation that ill-health is not desired, therefore research into ill-health is necessary;	Assumes as a foundation that science is a social good in itself; Assumes that science should always progress;
Assumes that a return to health, or the avoidance of ill-health, is to be achieved wherever possible; Applies ethical scrutiny to research with these pre-conditions in mind;	Assumes that scientists can or should 'go where the science takes them'; Generally assumes that scientific responsibility ends at publication; Tends to judge science in terms of what scientists think is important;
Is responsive to current health trends and concerns, often in the public eye;	Has to work within the law but often pushes the boundaries (not wrong in itself);
It is often assumed that health care research criteria are the best standards with which to assess other sorts of research;	Tends to view itself as at least one step removed from the public; Benefits to the public are not always a primary aim of the research (it may be theoretical);
The public can understand most health research in terms of why it is being done and what it is for;	Often does not see the need for people-based ethics in the way that social science does; Does not always specifically act to bring benefit (except to the scientist?);
Aims are set out clearly.	The public often do not understand why some research is being done or what it is for; Is this right?

Figure 7.1: Public and professional perceived differences and potential differences between a 'gold standard' ethics context and a science ethics context.

settings, or *vice versa*. Much NHS-related research is aimed at solving *recognised* problems, such as specific diseases, the need to improve procedures and investigation methods or to produce new medicines and vaccines and so on. The bottom line in most NHS-related research is that the approved research is taking place to provide imminent or soon-to-be achieved *specific* benefit for the human population. We usually *know* what the benefits will be or are likely to be.

In social science and in the natural sciences, while we want to guard the rights of participants, we are often doing research that *may or may not* have an application in the future. There may not be any pre-expected benefit as a result of the research. There may even be unexpected harm (obviously this can occur in gold-standard health settings too). We do not necessarily know where we are going beyond the end of 'this' project. This is very different from carrying out work to address a problem that we know already exists. The carrying out of research simply out of interest, or just to test an idea, is very different from attempting to solve an existing problem. Of course, ideally we want *all* sorts of research to go ahead. Many good, but unanticipated, outcomes have emerged from 'blue sky' research in the past and will no doubt continue to do so. The point is this — if we are at the forefront of open-ended research that is just 'out there looking to see where we can go', then we are, arguably, operating in a different philosophical field from the medical researcher who is trying to find a cure for a specific disease. Shouldn't we then be asking some extra ethical questions? This is where bio-chemical security ethics comes in.

We know that open-ended research journeys can in themselves be a part of the search for answers to specific problems, but much of this may only be recognised in hindsight. So yes, we want to do the exciting, new research, but we also need the ethics to keep up with it, if not to run ahead of it (ideally). In summary, what I am saying here is that just because something is prohibited or recommended in a gold-standard ethics approval process, this does not necessarily mean that it should be considered as prohibited or recommended in a biochemical security context.

As scientists we may need to apply extra, specific scrutiny that the 'gold standard' alongside our area either misses out (because the extra scrutiny is not needed in the field where the gold standard operates) or because our work may pose some additional ethical problems that would not be covered by the gold standard process itself. Just a thought.

How Do I Carry Out an Ethics Review of My Work?

All you need to do is run through The Toolkit questions again, focusing on any of the changes outlined below in the next section. You are unlikely to need to go over everything in great detail, but even focusing on an area of change (which usually triggers reviews) will help you to revise your position if necessary.

In order to simplify the review of The Toolkit questions, they are all repeated in an annex at the back of this book. You may be able to add some of your own questions to the list, based on the specific context in which you are working. It may be helpful to print the questions off and post them on the office or lab wall so that everyone can see them. This would help all the staff to feel involved with the ethics of the project.

You may like to have a standing ethics item on the agenda of your staff meetings and project meetings. By doing so, you will create an opportunity for anyone to raise an ethics issue without making a fuss over it. If opportunities are available to discuss concerns, people are more likely to engage with the issues. Such opportunities are also a good means through which to support junior members of staff, as they not only provide a forum for raising concerns or asking questions, but they provide learning opportunities in which juniors can observe seniors in ethical debate — and hopefully pick up some useful knowledge and skills themselves that they too will one day pass on. Given that most of the team will not have been involved in the initial ethics approval process, this is a useful way of getting everyone involved.

The point of review is the same as the point of initial approval:

- to avoid harming individuals through breaching their rights to autonomy,
- to take all possible precautions to minimise the risk of harm arising from your work, (including to your staff, others associated indirectly with your facility and the general public),
- by doing your best to promote beneficence where possible,
- by keeping up with professional standards and training through education and CPD,
- and by carefully considering your methods of dissemination bearing in mind your responsibilities under national and international law around the prohibition of biochemical weapons.

It is unlikely that you will need to reconsider everything. But do be aware that changes in one ethical area may trigger changes in another.

When Do I Need to Carry Out an Ethics Review of My Work?

All projects should be ethically reviewed at the start of planning the work, through the usual ethics approval processes and beyond.

Each project is different, but there are some general guidelines to help you decide when to revisit your applied ethics *after* you have received initial ethics approval. It is a safe option if you review the ethics status of your project at these points:

- Whenever anyone spots an ethical or bio-chemical security problem:
  - You should have procedures in place to enable staff to raise any problem, query or concern that they may have;
  - This must be open, safe and without repercussions to the person raising the issue;
  - You may need to pause the work while ethical and/or bio-chemical security review is carried out and discussed;
  - This does not necessarily mean that you have to go through full ethics approval again, but it may do — be prepared for this;
  - Maintain a good and open relationship with the relevant ethics board; if they know you are well-engaged with ethics they will be more disposed to help you find quick answers to your problems.

- Whenever there is a change to the research/work in terms of *scope*, *volume* or *nature*, e.g.
  - New working practices, new methods/ologies,
  - New materials,
  - New participants,
  - New researchers/assistants,
  - New workloads.

- When structural or procedural changes occur in the research/work place, e.g.:
  - Reconstruction, restructuring of hierarchies for the researcher (power relationships again),

    o Reconstruction, restructuring of hierarchies for the participants (ditto),

    o Having to relocate,

    o Having to deal with broken or old equipment,

    o New equipment or ways of doing the research or of what is being researched,

    o New staff or research participants,

    o Changed working hours or conditions.

- When there are any changes to your people, where they are, what they are doing, who they are working with, and so on …

- When unexpected events occur that may have relevance to risk management,

- When potential non-compliance with internal/external regulations is identified,

- When revising emergency response or contingency planning processes,

- As part of regular review of work practices (e.g. annually, monthly or more often as appropriate),

- Plus any other times when you recognise that there may be an ethical or bio-chemical security problem:

    o That has arisen unnoticed,

    o Has been flagged up by someone else who is not sure if there is a problem,

    o And you are not sure if a problem actually exists but you need to check it out and take advice.[1]

Do not panic — review is not as bad as it sounds. This list is just a guideline to help you think about the various events or processes that can and do occur in labs which trigger possible ethical and bio-chemical security issues. Once it is built in to all of your processes and you and your colleagues have become as accustomed to it as to biosafety processes, ethics and bio-chemical security awareness will just become part of your daily work.

---

[1] This list is adapted from materials produced by Sandia Laboratories, New Mexico.

The Research or Work Project: An Exemplar Plan of Action Using Applied Ethics

Identifying a possible research topic

- You may have a pre-set topic identified by a funding body seeking researchers to carry out the work,
  - o An ethics assessment will have been done (hopefully, or at least to some extent) but you should also carry out your own ethical review before accepting the work.
- You choose a topic that is of interest to you/and colleagues,
  - o You start by applying The Ethics Toolkit to the general area of interest as follows...

Staff and associated people

Who will be involved in the project? This is likely to be a long list....

- PI, research assistants,
- junior researchers, technicians, other laboratory staff,
- management staff of the facility,
- IT people,
- administrative staff,
- cleaning and maintenance staff,
- catering staff,
- delivery people, waste disposal people,
- others visiting or bringing people or materials to or from the facility,
- families and friends of these,
- others (identify),
- the wider public, locally, regionally and nationally.

Bear in mind that all of these people, and indirectly, those with whom they are in contact, will be associated with the work even when they are not on the premises where the work is carried out. They may carry pathogens or information with them deliberately or by accident to any place they go, at any time.

Where will they be located during and after the work?

- the laboratory,
- the wider facility premises,
- beyond the premises,
- homes, offices, shops, depots, storage facilities, postal premises,
- other open public places locally, regionally, nationally, and internationally,
- they may also have a cyber-presence that could be global in minutes (social media, email, blogs etc.)

When will they be associated with and/or working on the project?

- during working hours,
- during out-of-hours,
- on their vacations,
- when traveling away,
- at weekends and in evenings,
- in other words, all of the time.

What implications does this have for the physical location of notes, results, materials, equipment and so on during the project?

Materials or human sources of data

Who or what is the source of the project data?

- People
  - Can we access the relevant people without harming them?
  - Can we gather relevant data from them without harming them?
  - Can we involve them in the research/work in a meaningful way that will benefit them during the research/work? What about *after* the work?
  - What are our plans to inform them of the results of the research and to enable them to benefit from the results?
  - What role can they play in the formation of the research/work plans and the carrying out of the project?

    o What representation will they have on our project board/team?

    o What role will they have in making decisions during and after the project?

- Inanimate materials, pathogens or tissues

    o Can we acquire these in an ethical manner without harming the source(s)?

    o Are there any ownership issues that we need to address before, during or after the research/work?

    o What permissions do we need to acquire prior to accessing the materials?

    o From whom?

    o What processes and people will this involve? If necessary, when working with people as owners or sources, refer to the list above on 'People'.

    o Will the materials be returned to the original owners/source(s) after the research/work is complete?

    o What are our plans to store the materials during and after the research/work?

    o What are our plans to dispose of the materials during and after the research/work?

    o What actions do we need to take to ensure that the sources of the materials benefit from the research/work?

Now we can start to design the project ......

Identifying a Research Question

What question(s) can we ask of the data, bearing in mind the ethical rights and sensitivities of the sources?

People:
The Research Question must respect:

- The voluntary participation of

    o the sources of the data — the people,

    o the people involved from the workplace.

- The consent of

  o the people,
  o the people involved from the workplace,
  o people indirectly linked to the workplace (this may take a considerable amount of thought and discussion; do delivery drivers, for example, agree to transport whatever you need them to transport?).

- The privacy (personal security) of:

  o the people — anonymity or confidentiality, in time and in place,
  o the people involved from the workplace,
  o people indirectly linked to the workplace (there may be risks in being associated with your work, even indirectly).

The Research Questions must be such that it will be possible to give a scholarly, honest and open account of the research/work to the relevant audiences, bearing in mind the needs of biosafety and bio-chemical security.

The Research Questions must give rise to research/work that is reportable in an appropriate and acceptable way to relevant audiences.

The Research Questions must be such that the researchers, especially the PI, can accept responsibility for the outcomes as far as is reasonably foreseeable at the time of the research being carried out and completed.

The Research Question should not usually lead to actions causing un-agreed harm or damage or destruction to the sources of the materials — agreement must be reached with the people who are the source or owners wherever possible.

Planning a data generation strategy

Ask yourself these questions about the people who are supplying you with the data, or who are supplying you with the materials from which you will gather data.

- Are my data generation activities impeding or negating these ethical rights of the people who are the source of the data through issues to do with:

  o Their voluntary participation?
  o Their reasonably informed consent?
  o Their privacy — anonymity or confidentiality and security in time and space?
  o Confidentiality and security of work and materials?

- Ask yourself these questions about your ethical responsibilities as a researcher/worker:

  o Am I acting in such a way as to reduce, minimise or avoid harm for the donors involved and for the people involved from the workplace and beyond?
  o Am I at least attempting to offer some benefit for the people who are the source of the data, or who are providing materials from which I will collect data?
  o Am I practising good scholarship skills in my notes, report-writing and other forms of accounting for the data I am generating?
  o Am I reporting my work correctly, in a timely manner and to the right audiences/authorities?
  o Am I prepared to stand by my data generation methods if they are criticised and justify them ethically and methodologically?
  o Am I prepared to take responsibility to justify the outcomes of my data generation methods?
  o Am I satisfied, during the research or work project, that I am meeting my responsibilities under the BTWC and the CWC by doing my best to ensure that my work is protected, to the best of my ability, from misuse?

Publication and post-publication stages

What *reasonable* steps do we need to take to ensure that publication of the work does not cause harm to people/animals/plant life now or in the *reasonably foreseeable* future?

The answer to this question depends on the nature and scope of your work, your skills as a scientist or person having oversight of scientific work, the skills of those involved in the work, their personal and professional approaches to the work, and the potential applications to which the work may be put.

- Have we considered how the work may affect people in different geographical areas as well as at different times?
- Do we need to consider redacting some information in reports, papers or presentations?
- How?
- Why?
- Will the redactions need to be the same in all publications, or just some?
- Will a time come when the redactions could be opened to allow full open access to the work?
- What other outlets are open to us to disseminate the work to those who need access to it?
   - How will we know who needs access to it?
   - How can we arrange suitable and appropriate ways to share our work with those who we trust?
   - How do we know we can trust them?
   - What steps, if any, can we take to keep track of what is being done with our published *or unpublished* work?

This chapter may have opened up some new ways of thinking for you. It may have challenged you or put you off ethics even more. However, while this all seems very daunting and time-consuming when it is new, just consider again how easily you factor biosafety reviews and activities into your daily work within projects already. If you were new to biosafety and unfamiliar with the principles and application of it, you would feel daunted by that too. But because you do understand biosafety in principle and in action, you accommodate it easily into your everyday work.

Applied bio-chemical security ethics is the same. Once you have become familiar with it and have applied it in practice, it too will become just another strand of your daily work. Ethical problems will be easy to recognise and deal with. All you need to do is to make a decision to factor this into all of your planning, work and review processes and it will become as easy as biosafety. Honestly.

Chapter 8

Still Not Convinced that You Need to Address Bio-Chemical Security in Your Work?

This chapter contains some examples of real-life research or activities of concern that had dual use security implications. They are all drawn from the United States of America and the United Kingdom. I have chosen to look only at these two countries for a reason. Typically, the US and the UK are held up as examples of 'how to do biosafety properly'. Yet, as you will see in the following pages, the US and the UK still have major problems with biosafety. What are the chances, therefore, that some other countries with less well developed policies can do any better? There are many more real-life examples if you look for them, but this selection is included simply to illustrate how 'good' science can still pose a risk to the public. They all involve activities, intentional and unintentional, that provided opportunities for pathogens or information to be acquired by people who could have misused these for hostile purposes.

In some cases this could have arisen through accidents and in others through negligence or unauthorised or ill-advised activities. Most of these cases also illustrate problems that arise from an over-reliance on the efficiency and effectiveness of existing biosafety procedures, including the oversight of these. Remember also that even *mistakes* in standard biosafety procedures can lead to biochemical 'misuse' by accident or unintentionally. In order to achieve effective bio-chemical security we need to have effective biosafety. If the biosafety is compromised, then effective bio-chemical security cannot be achieved. You will find that many scientists engaged in controversial work tend to use the 'effective biosafety' argument as means of effecting 'biosecurity'. You may agree with them. Read this chapter and see if, by the end of it, you still believe that biosafety can be relied upon as a means of achieving bio-chemical security.

I have included a range of UK and US examples here. This is not because these countries have a bad record in ethics, but because they are arguably better at admitting their mistakes (although there is still much room for improvement). Bear in mind that if the following problems can occur in well-regulated science activities, what sort of problems are likely to occur in less-regulated countries?

The Pirbright Foot and Mouth Outbreak 2007

Pirbright in Surrey, England, is the location of the Pirbright Institute (formerly the Institute for Animal Health — IAH), a world-recognised site of expertise in the research and surveillance of virus diseases of farm animals and viruses that spread from animals to humans.[1] The site contains a range of high containment laboratories, reference laboratories, and research facilities; it carries out contract work and much significant research. Like all working laboratories, it is subject occasionally to unwanted occurrences.

In early August 2007 an outbreak of Foot and Mouth Disease (FMD) was identified by DEFRA[2] at a farm close to the Pirbright site. A few days later an outbreak was confirmed at a second nearby farm. Both infection sites farmed beef cattle. Given the severity and impact of the FMD outbreak in 2001 in the UK, this became a big news story and was the subject of much media attention. FMD is a severe,

---

[1] Pirbright Institute website. Available on: http://www.pirbright.ac.uk/About/about. aspx. Accessed 16/12/15.
[2] DEFRA is the Department for Environment, Food and Rural Affairs, UK Government.

highly communicable viral disease of cloven-hoofed animals (cattle, swine, sheep, and goats) and can also affect a variety of wild animals, including deer. It presents a serious risk to farm livestock and is subject to statutory control measures, which include emergency procedures to contain and eradicate the disease.[3]

The Health and Safety Executive were called in, with other experts, to investigate biosecurity at the Pirbright site and to produce a report for the government. The investigation found that the likely exposure time of the cattle had been in late July 2007 and that the second farm was probably infected through contact with the first farm. The strain of FMD involved was identified by DEFRA as Type 01 BFS67. This strain, while being involved in an FMD epidemic in the UK in 1967, was not known to be in circulation anywhere in the world at the time of the 2007 outbreak. The only site at which it was held in the UK at that time was Pirbright, which was only 4.6 km from the first infected site and 2.9 km from the second infected site. The report found that:

- Such was the condition in which we found the site drainage system [at Pirbright] that we conclude that the requirements for Containment Level 4 were not met, thus constituting a breach of biosecurity for the Pirbright site as a whole. Our conclusion is supported by the evidence we found of long-term damage and leakage, including cracked pipes, unsealed manholes and tree root ingress. We have investigated ownership of the drainage system, which rests with IAH..... [IAH is the Institute of Animal Health, the former name of the Pirbright Institute]... Moreover, we judge the practice employed by IAH of using bowsers and hoses in the intermediate site effluent drains to clear blockages without a standard operating procedure (SOP) to be a breach of biosecurity.
- We established that, in the period covered by our investigation, not all human and vehicle movements via the IAH gatehouse to

---

[3] Health and Safety Executive (2007) Final report on potential breaches of biosecurity at the Pirbright site. 2007, p. 7. Available on: http://www.hse.gov.uk/news/2007/finalreport.pdf. Accessed 16/12/15.

the site were recorded, in particular traffic associated with construction work going on at the site at the time. We conclude these failures to keep complete records were not in line with accepted biosecurity practice and represent a breach in biosecurity at the IAH site. We also found evidence of poor monitoring and control of access to restricted areas within IAH facilities. We conclude that this too constitutes a breach in biosecurity.

- We judge it likely that waste water containing the live virus strain, having entered the drainage pipework, then leaked out and contaminated the surrounding soil. We also believe that excessive rainfall may have exacerbated the potential release from the drain.

- We conclude it likely that those vehicles, having driven over this part of the site, carried off out of the site materials containing the live virus in the form of mud on tyres and vehicle underbodies.

- We have further established that some of the vehicles thus contaminated drove from the site and along Westwood Lane, Normandy... [which] passes the first infected farm. It is our conclusion that this combination of events is the likely link between the release of the live virus from the Pirbright site and the first outbreak of FMD.[4]

In other words, the virus was likely to have leaked out of the drainage system, been picked up by vehicles and deposited outside the site near the first infected farm. It infected cattle at the first farm, and then probably passed to the second infected farm from there.

This is an example of how any high quality facility may find itself in with an unexpected major incident. We can all express amazement at this situation in hindsight, but if issues such as this can arise at a site like Pirbright, what can you say about your facility? This could happen anywhere at any time. Pirbright was and is a world-leading science facility that was subject to all the usual oversight but it still found itself with a major problem. This case also illustrates the potentially

---

[4] Health and Safety Executive (2007) Final report on potential breaches of biosecurity at the Pirbright site 2007, p. 3. Available on: http://www.hse.gov.uk/news/2007/finalreport.pdf. Accessed 16/12/15.

confusing use of terminology that I mentioned in an earlier chapter. Here, the term 'biosecurity' is used in the biosafety sense of the word — meaning security in terms of containment of a pathogen, in this case a strain of the FMD virus. This meaning of 'biosecurity' is often used in the UK in relation to agriculture and farming. However, this situation did pose a biosecurity risk in terms of what we have been calling bio-chemical security in previous chapters.

Because of unrecognised failings in procedure and infrastructure at Pirbright, the FMD virus was, in effect, made available outside controlled laboratory conditions *by accident*. How easy would it have been for an ill-intentioned person to acquire a sample during the period after the leak was identified so that the virus could be used later for hostile purposes? The other obvious issue raised by this case is that security problems inherent in infrastructure offer useful access to all sorts of 'waste' that could be accessed by those who would misuse it. Here, it was a problem with drainage and waste water, but it could equally, in another setting, involve a failing autoclave or inadequate waste disposal methods (common in many countries).

The 2001 FMD outbreak cost the UK economy an estimated £3 billion for the public sector and £5 billion for the private sector.[5] This incident in 2007 caused a high degree of concern amongst the agricultural community and the government. What price the effect of a small illegally obtained sample as a biological weapon?

The Thomas Butler Case 2003

This example involved the prosecution of a well-known scientist based at Texas Tech in Lubbock, Texas — Dr. Thomas Butler. The case raised massive reaction both against Butler and in his support. If you search online for 'Thomas Butler prosecution' you will find a substantial amount of information about the case.[6] I am not going to get into

---

[5] National Audit Office's final report on the UK's. *2001 Outbreak of Foot and Mouth Disease*. Available on: https://www.nao.org.uk/press-releases/the-2001-outbreak-of-foot-and-mouth-disease-2/. Accessed 16/12/15.

[6] A good place to start is the Federation of American Scientists website. Available on: http://fas.org/programs/bio/factsheets/thomasbutler.html.

the rights and wrongs of the prosecution, or of the outcomes, but I will mention some ethical issues that arose during the events that led to Butler's prosecution — all based on Butler's choices when faced with various courses of action.

Thomas Butler was an expert on plague, which is still a health issue in the US and in many countries around the world.[7] He had a long professional history of work in the US and overseas, including much research into infectious diseases. From 1987, he worked at Texas Tech University as Chief of the Infectious Diseases Department, a position he held until he was forced to resign in January 2004.

In 2000 Butler began a collaboration with scientists in Tanzania, looking at ways to improve antibiotic therapies against *Yersinia pestis* (the plague-causing agent). In January 2003 Butler reported to his biosafety officer at Texas Tech that he could not locate 30 vials of plague specimens. As a result of this, a large number of FBI agents (60+) arrived at the Texas Tech site and began an investigation. Butler, after long questioning by FBI agents, signed a document saying that he had possibly destroyed the missing samples by autoclaving them. He apparently underwent interrogation and signed this document in the belief that it would prevent legal action against him.

It did not. He was detained and prosecuted. A number of other charges were eventually made, including the illegal importation, smuggling, exportation, and transportation of hazardous materials based on his conduct in transporting and transferring the plague bacteria, lying to the FBI, misappropriation of funds and some tax issues (69 charges in total). Butler was duly convicted of some of the charges. He appealed but failed to have the convictions overturned. He was jailed for two years and had to pay a fine. This caused uproar in the science community.

It emerged in the investigation that Butler had illegally transported plague samples in his personal luggage on various international flights (British Airways and American Airlines) as he returned from Tanzania to the US, then on internal flights in the US as well,

---

[7] CDC website. Available on: http://www.cdc.gov/plague/maps/ *Plague in the United States*. Accessed 16/12/15.

in April 2002. He apparently did not have the relevant permits or clearance to transport these samples, did not declare them to customs and failed to package and label them as required by US law. He was also charged with illegally transporting samples from Lubbock, Texas to the Centers for Disease Control and Prevention in Fort Collins, Colorado in his private vehicle, without the necessary CDC permit for the transportation of *Y.pestis*.

In relation to the shipping of 30 vials of *Y.pestis* back to Tanzania by the Fed-Ex postal service, Butler failed to properly identify the specimens inside the package, did not attach a hazardous materials label and sent them without an export licence in September 2002. It was claimed that approval for this was required by both the US and Tanzanian authorities. It was also claimed that in sending this package to Tanzania, Butler filed a false export control document describing the package as 'laboratory materials' when he knew that it contained a government-regulated select agent (one subject to strict governmental oversight and controls).

Butler was acquitted of the charges relating to the transportation of the samples from Tanzania to the US (the jury apparently believed that he has acted in good faith at the time), but was convicted of the postal incident back to Tanzania among other outcomes. What do you think of this?

Without commenting on the legal issues or the rightness or wrongness of the prosecution and its outcomes, this case raises many ethical issues. Let us start with autonomy and consent. If you buy an air ticket to travel overseas or within your own country, do you expect to be sitting near someone who has an undeclared deadly agent in their personal luggage? Would you actively choose to be a fellow-passenger with Dr. Butler on those flights between Tanzania and the US if you were told in advance about the presence of unlicensed and improperly packaged and identified plague samples in the hold? I would not. The same applies to the internal flight to Lubbock. Accidents happen. What if one of the planes had crashed? Is not there enough to do for emergency services and possibly survivors without being faced unknowingly with the plague? What if the sample containers were broken during the journey? Can we be absolutely certain that Butler was not himself contaminated with the plague pathogen during this journey? The same issues arise in the period when Butler drove samples in his car from Texas to Colorado. What if he had been involved in a road accident and the packaging was breached?

The posting of plague samples back to Tanzania was done via Fed-Ex, with inaccurate or misleading labels that failed to identify the true nature of the contents of the package. Much argument followed about what was legal at the time and so on. But the fact remains that Butler knowingly put plague samples in the post without proper labelling or the required permits. What about the rights to safety and self-determination of all the postal workers down the chain back to Tanzania? And all the people who came into contact with them? Again, what if the packaging or containers were breached? People would not know what they had been exposed to and would likely have become ill and possibly have even died. Where is the consent for that? Where is the right to safety of people in contact with the materials?

These problems can be drawn together under the umbrella of 'deception.' It does not matter in practice whether deliberate deception was *intended*, because deception *occurred* as a result of Butler's actions. Does it make any difference if he thought he was acting in good faith at the time? Was Butler entitled to assume the consent of his fellow passengers and the postal workers to travel alongside his dangerous package of materials? Did he even *think* about the ethical issues of what he was doing? I do not know.

The problems do not only apply to what Butler did. When he rightly declared his Tanzanian project to Texas Tech, the institutional board there decided there was no need to carry out a full ethics review of the project as Butler was involved as a consultant without patient care responsibilities. That decision was obviously considered safe at the time, but what did Texas Tech look like later when all this came to the attention of the public?

Much of the scientific defence of Butler since the case emerged has focused on the wonderful work that he had done prior to this incident. Many prominent institutions and individuals, including Nobel Laureates, spoke out on his behalf, criticising the government and the FBI. Jonathan Turley, a George Washington University law professor who represented Butler in his unsuccessful appeals, was quoted in the US press as saying that Butler had spent his life in protecting and healing people and that 'He would never do anything that would endanger people.'[8] This may well be true, but was he not endangering people without their knowledge in transporting plague samples illegally? Is not that 'danger'? It is easy after the event to say that nothing went wrong. But what if something *had* gone wrong?

Thomas Butler lost his job and his licence to practice. His career was ruined. His passport was seized for a time. Another scientist wrote in 2006, after Butler's release from prison, that Roosevelt's quote about having nothing to fear except fear itself could now be replaced with 'we now have nothing to fear except misdirected ignorance and panic in relation to bioterrorism.'[9] Do you agree? It is difficult to reach a cut and dried decision on this from an ethics point of view.

On the one hand, Butler's work has been influential and effective; he was engaged in this work for the common good. This is ethically admirable. On the other hand, he failed to follow regulations that were in place for good reason, and thereby exposed others to (unreasonable) risk. This is ethically problematic. Does Butler's previously good work somehow nullify his later ill-advised actions?

Synthesis of the 1918 Flu Virus

In 2005, a team of scientists led by Jeffery Taubenberger of the Armed Forces Institute of Pathology, Washington, DC, decided to reconstruct the pandemic 1918 influenza virus in order to study the

---

[8] Associated Press (2010) Professor in 2003 plague scare sets off Miami airport shutdown with canister. Available on: http://www.cleveland.com/nation/index.ssf/2010/09/professor_in_2003_plague_scare.html. Accessed 16/12/15.

[9] Greenhough, W.B. (2006) Update on Dr. Thomas Butler, *Clinical Infectious Diseases* 43, 259–260. Available on: http://cid.oxfordjournals.org/content/43/2/259.full.pdf. Accessed 16/12/15.

properties associated with its extraordinary virulence.[10] This was a matter of choice on the part of the team. They recovered genomic RNA of the 1918 virus from archived autopsy materials and from the lung tissues of a flu victim who had been buried in permafrost in 1918. The coding sequences were determined and analysed, providing information about the nature and origin of the virus. During the research all the viruses with gene segments from the 1918 virus were worked on under BSL3 laboratory conditions in accordance with relevant national guidelines. Further, the staff involved in the work were also taking antiviral prophylaxis as well as adhering to strict biosafety regulations.

Let us think about this from a security perspective. The original 1918 influenza pandemic killed between 20 million and 50 million people worldwide (some say more). Prior to carrying out the reconstruction of the virus, Taubenberger and his team had developed useful methods of analysis for investigating preserved human tissues. They had already used their techniques to investigate colour blindness in a preserved human eye, and decided to look for something else on which to try out their ideas.

Taubenberger said 'The 1918 flu was by far and away the most interesting thing we could think of because it's not just a historical curiosity... The 1918 pandemic was the most lethal infectious disease outbreak probably in all history and no one was able to study it because the virus was not isolated at the time — influenza viruses were not even known to exist in 1918.'[11]

At the time of the work, two types of antiviral drugs, rimantadine (Flumadine) and oseltamivir (Tamiflu), were already known to be effective against viruses similar to the 1918 virus. Vaccines containing

---

[10] Tumpey, T.M., Basler, C.F., Aguilar, P.V., Zeng, H., Solozarno, A., Swayne, D.E., Cox, N.J., Katz, J.M., Taubenberger, J.K., Palese, P. and Garcia-Sastre, A. (2005) Characterization of the Reconstructed 1918 Spanish Influenza Pandemic Virus *Science* 310, 77–80. Available on: https://www.sciencemag.org/content/310/5745/77.abstract. Accessed 30/12/15.

[11] American Society for Microbiology (2006) website — Meet the Scientists: Jeffery Taubenberger. *Mining History to Fight the Flu*. Available on: http://archives.microbeworld.org/scientists/interviews/interview10.aspx. Accessed 17/12/15.

the 1918 HA or other subtype H1 HA proteins were also known to protect mice against the 1918 virus and the current influenza vaccine provided some level of protection against the 1918 virus in mice. Bearing this in mind, many non-scientists would ask 'Why was this work done at all?'

Given the dangerous nature of both polio and 'Spanish flu' virus, WHO attempts to eradicate wild polio virus and the successful eradication of the 1918 flu virus, were these experiments ethically and scientifically justified? Were they necessary? What dual-use concerns did they raise? Do the benefits of such research outweigh the costs? What other frameworks or mechanisms could be used to reach a decision on the ethics or value of this work?

Much emphasis in reviews and the initial publication was placed on the stringent biosafety measures taken during the research. The researchers, in their paper in *Science* in 2005, also cited work by other authorities emphasising the high likelihood of a future influenza pandemic that could be equally virulent to the 1918 virus. An information page on the website of the Centers for Disease Control and Prevention (US)[12] focusing on this research, seeks to allay public concerns on a variety of fronts:

> The work described in this report was done using stringent biosafety and biosecurity precautions that are designed to protect workers and the public from possible exposure to this virus (for example, from accidental release of the virus into the environment).

> Before the experiments were begun, two tiers of internal CDC approval were conducted: an Institutional Biosafety Committee review and an Animal Care and Use Committee review.

> Biosafety Level 3 or Animal Biosafety Level 3 practices, procedures and facilities, plus enhancements that include special procedures ..... are recommended for work with the 1918 strain.

---

[12] Centers for Disease Control and Prevention website (updated 2014) *Reconstruction of the 1918 Influenza Pandemic Virus: Questions & Answers*. Available on: http://www.cdc.gov/flu/about/qa/1918flupandemic.htm.

A Biosafety Level 3 facility with specific enhancements includes primary (safety cabinets, isolation chambers, gloves, and gowns) and secondary (facility construction, HEPA filtration treatment of exhaust air) barriers to protect laboratory workers and the public from accidental exposure. The specific additional ("enhanced") procedures used for work with the 1918 strain include:

- Rigorous adherence to additional respiratory protection and clothing change protocols;
- Use of negative pressure, HEPA-filtered respirators or positive air-purifying respirators (PAPRs);
- Use of HEPA filtration for treatment of exhaust air; and
- Amendment of personnel practices to include personal showers prior to exiting the laboratory;
- Highly trained laboratorians can work with the 1918 influenza virus strain safely using BSL-3-enhanced containment. Researchers at CDC have specialized training and go through a rigorous biosafety (and security) clearance process. For the work reported in the *Science* article, the lead CDC researcher provided routine weekly written reports to CDC management officials, including the agency's Chief Science Officer, and was instructed to notify agency officials immediately of any concerns related to biosafety or biosecurity.

Does the Generation of the 1918 Spanish Influenza Pandemic Virus Containing the Complete Coding Sequence of the Eight Viral Gene Segments Violate the Biological Weapons Convention (BWC)?

No. Article I of the BWC specifically allows for microbiological research for prophylactic, protective, or other peaceful purposes. Article X of the BWC encourages the "fullest possible exchange of... scientific and technological information" for the use of biological agents for the prevention of disease and other peaceful purposes. Further, Article X of the BWC provides that the BWC should not hamper technological development in the field of peaceful bacteriological activities. Because the emergence of another pandemic virus is considered likely, if not inevitable, characterisation of the 1918 virus may enable us to recognise the potential threat posed by new influenza virus strains, and it will shed light on the prophylactic and therapeutic countermeasures that will be needed to control pandemic viruses.

Does the Report Provide a "Blueprint" for Bioterrorists to Develop and Unleash a Devastating Pandemic on the World?

No. This report does not provide the blueprint for bioterrorists to develop a pandemic influenza strain. The reverse genetics system that was used to generate the 1918 virus is a widely used laboratory technique. While there are concerns that this approach could potentially be misused for purposes of bioterrorism, there are also clear and significant potential benefits of sharing this information with the scientific community: namely, facilitating the development of effective interventions, thereby strengthening public health and national security.

This webpage is interesting because it clearly exists to address public concerns, which is admirable. However, the comment that the research does not violate the BTWC is not as straightforward as it seems. Work has been going on for many years by a range of institutions and groups to highlight the biosecurity (or bio-chemical security) risks that arise from the close overlap between 'prophylactic, protective, or other peaceful purposes' and the misuse of the same science for hostile purposes under the BTWC. The very act of carrying out certain types of work, even though for 'prophylactic, protective, or other peaceful purposes' is in itself an act that carries increased risk of potential misuse (if a pathogen does not exist, it cannot be misused). Acceptance of these risks is disputed by many scientists, often citing medical justifications, academic and scientific freedom and the protections afforded by stringent biosafety procedures. But as we have seen in earlier chapters, there is a very fine line between the offensive and defensive potential of such research.

Do you think that the benefits of this research were worth the security risks of reconstructing this virus? Treatments for similar viruses were already known to be effective, and the current vaccine was also known to be protective in mice (work which could be developed further). Were there other ways to enhance our understanding of the virulence of pandemic viruses rather than reconstructing this deadly one?

In relation to the *publication* of the research, the CDC webpage seems to claim that the full publication of the work was not a security risk simply because the technique used was already widely known. But had it ever been applied to something as potentially dangerous as this before? Some attempt is made in the next sentence at a 'cost-benefit' decision but clearly comes down on the side of the 'benefits' at the expense of the 'costs' (as it always does). What decision-making process was used to arrive at this decision? Who made this decision? Were there any dissenters at the time? If so, what happened to their views?

According to some authorities, no other global epidemic has claimed as many lives as did the influenza pandemic of 1918; some estimates put the figure at around 70 million individuals.[13] Did we need to reconstruct a virus that killed up to 70 million people? I do not know the answer to this, but I do wonder at some of the justifications behind this work. What do you think about Taubenberger's comment that 'The 1918 flu was by far and away the most interesting thing we could think of...'? Is that a good reason to go ahead and do this? You can read an interesting interview with Jeffrey Taubenberger online, in which he discusses his work with the 1918 virus. See what you think.[14]

Do you agree that in the current global security climate we need to take more notice of these and similar security risks? A quick glance at the evening news and other daily media clearly illustrates the increasing sophistication of those who would seek to take hostile actions against the peaceful societies of the world.

Let us look at some comments that have been made since this work was done on the reconstruction of the 1918 flu virus. At the

---

[13] College of Physicians of Philadelphia (2011) History of Vaccines blog: *Spanish Influenza Pandemic and Vaccines*. Available on: http://www.historyofvaccines.org/content/blog/spanish-influenza-pandemic-and-vaccines.

[14] Conversations with Pathologists (2007) Jeffrey Taubenberger. Available on: http://www.pathsoc.org/conversations/index.php?view=article&catid=65%3Ajeffery-taubenberger&id=92%3Ajeffery-taubenberger-full-transcript&option=com_content. Accessed 17/12/15.

time of publication of the study, Phillip A. Sharp, Institute Professor at the Massachusetts Institute of Technology said:

> 'I firmly believe that allowing the publication of this information was the correct decision in terms of both national security and public health.'[15]

Further comment came from The Sunshine Project, a Non-Governmental Organisation with offices in the US and Germany. The organisation closed in 2008 but maintained its website for some years after this. It worked 'against the hostile use of biotechnology in the post-Cold War era, researching and publishing to strengthen the global consensus against biological warfare and to ensure that international treaties effectively prevent development and use of biological weapons.'[16] In 2003 when it became known that reconstruction of the virus was going ahead, Jan van Aken, a former biological weapons inspector for the UN, German politician, Greenpeace activist and biologist with The Sunshine Project, said:

> 'It simply does not make any scientific sense to create a new threat just to develop new countermeasures against it......Genetic characterisation of influenza strains has important biomedical applications. But it is not justifiable to recreate this particularly dangerous eradicated strain that could wreak havoc if released, deliberately or accidentally....If Jeffery Taubenberger worked in a Chinese, Russian or Iranian laboratory, his work might well be seen as the 'smoking gun' of an offensive biowarfare program...'[17]

---

[15] Sharp, P.A. (2005) Editorial: 1918 Flu and Responsible Science. *Science* 310 (5745), 17. Available on: http://www.sciencemag.org/content/310/5745/17.full. pdf. Accessed 18/12/15.

[16] Federation of American Scientists (2010) *The Sunshine Project* Virtual Biosecurity Center website. Available on: http://www.virtualbiosecuritycenter.org/organizations/the-sunshine-project. Accessed 18/12/15.

[17] Van Aken, J. (2003) A Sunshine Project News Release. Formerly available at the Sunshine Project website. www.sunshine-project.org. The English language site is currently unavailable but the German site remains online at http://www.sunshine-project.de/. Accessed 18/12/15. The news release is currently copied on several

Among many other findings using Freedom of Information requests, the Sunshine Project found that the US Department of Agriculture (USDA) and the University of Georgia, where work on the 1918 flu virus reconstruction was carried out, failed to convene a meeting of the relevant Institutional Biosafety Committee to review the plans for the work. Through Freedom of Information (FOI) requests, the Sunshine Project found that the university signed off the work based on informal discussions with only four members of the committee. The lack of a formal meeting appears to have led to no formal opportunity for concerns to be heard.[18]

Since this period, security concerns have moved on even further and there is arguably greater awareness of the potential problems with work such as this. However, the 'cost-benefit' argument apparently remains the justification of choice in many cases and debates. This is not the place for a lengthy debate on the 'for' and 'against' arguments in such instance, but having read the previous chapters on applied ethics, do you think differently now about situations like this? If not, what argument are you using apart from the 'cost-benefit' assessment? If you are committed to the cost-benefit approach, what decision-making process are you going through? Are you happy to rely on the stringent application of biosafety measures to protect the public from exposure to recreated or modified pathogens?

The H5N1 Controversy

In late 2011 it came to the attention of the public that two teams of researchers (one in the Netherlands under Ron Fouchier at Erasmus Medical Center in Rotterdam and the other under Yoshihiro Kawaoka at the University of Wisconsin, Madison) had submitted papers to

---

sites including that of Green Left Weekly (an Australian activist site) at www.green-left.org.au/node/28862. Accessed 18/12/15.

[18] Sunshine Project (2003) *Biosafety Irregularity In Spanish Flu Experiments: Highlights The Need to Strengthen Biodefense Transparency*. Available on: www.sunshine-project.org/publications/pr/pr211003.html (no longer available, but cited by PBS at http://www.pbs.org/wgbh/nova/sciencenow/3318/02-poll-sources.html. Accessed 18/12/15.

prestigious journals (*Science* and *Nature*) describing their work on the mammalian transmissibility of an H5N1 avian influenza strain. Considerable public and scientific debate followed this news, particularly since this strain of the influenza virus had significant pandemic potential. Professional scientists and the public questioned the suitability of the research and raised concerns about the publication of the details of the work as a security and public health risk. One of the main areas of argument that followed centred on the effective 'censorship' of science implied in any demands to withhold publication or require redactions, with the rights of scientists to carry out and publish controversial work freely being balanced (or not) with the rights of the public to be protected from highly dangerous 'invented' pathogens developed in the laboratory. So what had the two teams actually done?

They had used existing and well-known techniques to identify and sequence genetic mutations of the H5N1 influenza strain; they then inserted the mutated genes into a new virus (H1N1) and were able to establish respiratory transmission of influenza between ferrets. This was contentious because it showed that relatively few genetic changes in H5N1 viruses enabled transmission via the respiratory route in these animals, which are used in influenza work because their response to influenza is considered to be very similar to that of humans. In other words, the researchers enhanced the transmissibility of an existing influenza virus by manipulating it in a way that enhanced its likely virulency amongst human populations.

Much debate was had in the scientific and the general media. Some of the argument focused on the work itself and some on the publication of the work. Many people (including some scientists) were worried about the risks of accidental release and the possibility of misuse of the work by those with ill intent. Some argued that the work should not be published, some that the work should not have been done in the first place. The lead scientists of the two teams, Fouchier and Kawaoka, argued for their freedom to publish as scientists and said that their work was a vital plank in the construction of future pandemic preparedness (our old friend the 'cost-benefit' approach again).

Remember the Fink Report of 2004? You will recall that as a result of this report, the US National Science Advisory Board for Biosecurity (NSABB) had been set up to look into science of concern to the security community. The NSABB committee duly considered the two H5N1 papers and concluded, in December 2011, that the papers should be only partially published, leaving out details of the experimental design.

A series of events followed, which may be summarised as follows:

- In January 2012 the two research teams agreed on a temporary moratorium on their research around H5N1 strains,
- In February 2012 the World Health Organization (WHO) held a meeting with a range of influenza and public health experts and recommended that the papers both be published *after* the biosecurity concerns were addressed and a communications plan to 'increase public awareness and understanding of the significance of these studies and the rationale for their publication' had been devised,[19]
  - The WHO noted that the studies of both teams were 'essentially proof-of-principle experiments' which had not been designed to clarify pathogenicity or the degree of transmissibility of lab-modified viruses;
  - The WHO also noted that there was a significant public health argument to the full publication of both papers, but recognised that some serious social concerns were also raised by the work;
  - It recommended that 'a review of the essential biosafety and biosecurity aspects of the newly developed knowledge' be undertaken prior to publication.

- In late March 2012, a new US policy for the oversight of life sciences and 'dual use research of concern' (DURC) was published by the Office of Biotechnology Activities (US Dept of Health and Human Services),

---

[19] World Health Organisation (2012) *Report on technical consultation on H5N1 research issues*. Geneva: WHO. Available on: http://www.who.int/influenza/human_animal_interface/consensus_points/en/. Accessed 28/12/15.

- Also in late March 2012, the NSABB committee reconsidered the *revised* manuscripts and voted (not unanimously — 6 out of 18 members disagreed) in favour of the publication of the revised papers *in full*,[20]
- 2 May 2012, publication of the Kawaoka paper,[21]
- 22 June 2012, publication of the Fouchier paper.[22]

What is your opinion of all of this? Was the work as necessary for future pandemic preparedness as it was claimed to be? I am not an expert on the influenza virus so I do not know. However, what about the arguments around 'scientific freedom' and the heavy reliance on existing biosafety measures? I would argue that we do not have scientific or academic freedom *already*.

Can we do any research we like, or experiment on humans in any way that 'the science takes us'? No, we are restricted by socially-acceptable and debated norms that dictate what we can and cannot do in terms of research using human participants or human tissues. Since the horrors of the Second World War and the horrific 'medical' experiments carried out by the Germans and the Japanese, we have formulated norms to prevent any repeat. We have laws in place to influence and regulate scientific research. In years of training PhD candidates I have repeatedly had to emphasise to students that our norms of autonomy (self-determination) and the value of human

---

[20] National Science Advisory Board for Biosecurity (2012) *Findings and Recommendations March 29–30*. Available on: http://osp.od.nih.gov/sites/default/files/resources/03302012_NSABB_Recommendations_1.pdf. Accessed 19/12/15.
[21] Imai M., Watanabe T., Hatta M., Das S. C., Ozawa M., Shinya K., Zhong G., Hanson A., Katsura H., Watanabe S., Li C., Kawakami E., Yamada S., Kiso M., Suzuki Y., Maher E. A., Neumann G. and Kawaoka Y. (2012) Experimental adaptation of an influenza H5 HA confers respiratory droplet transmission to a reassortant H5 HA/H1N1 virus in ferrets, *Nature* 486, 420–429.
[22] Herfst S., Schrauwen E.J., Linster M., Chutinimitkul S., de Wit E., Munster V. J., Sorrell E. M., Bestebroer T. M., Burke D. F., Smith D. J., Rimmelzwaan G. F., Osterhaus A. D. and Fouchier R. A. (2012) Airborne transmission of influenza A/H5N1 virus between ferrets. *Science* 336(6088), 1534–1541.

rights prevent us from misleading or harming humans in any way — in social science and in the natural sciences. Of course, these norms vary from culture to culture, but the general global acceptance of 'western' standards is largely respected, in principle even if not always in action.

Can we experiment in any way we choose on animals? No. Scientific work on animals is highly regulated and subject to stringent oversight. Obviously this is an area of great social contention and not an issue that can easily be endorsed. However, we are essentially heavily restricted in what we can 'do' to animals in the name of research — which is as it should be.

Do we already fully publish everything that we do in research? No. There are many reasons why some research is not fully published, or even not published at all. Work carried out for military or other governmental purposes is often subject to contracts forbidding publication. The other area in which full publication is not always pursued is that of commercial gain. It is common practice to apply for patents when scientific research produces new technology or other new initiatives that could provide a financial gain for the 'inventors'. Why do we not hear scientists shouting about 'scientific freedom' and the rights of other scientists to have full access to everything there?

I believe that we need to have a new debate about both rights and responsibilities in this area. We cannot live in a society where the rights of all can be assessed and prioritised by a few with a vested interest in the short-term outcomes of their decisions. I am *not* promoting the censorship of science — although we already have that in effect. In terms of commercial gain, we already operate a 'pick and mix' approach when we want to be careful about publicising 'our' science if we think we may make some money from it. I *am* in favour of a more highly-developed social awareness and accountability among the science community. This will involve scientists and all the associated policy makers, managers and 'interested parties' getting together, *with lay people* on a regular *formal* basis to debate the issues and to make open and transparent decisions about difficult work through an open and transparent decision-making process such as applied ethics. All 'sides' need the opportunity to have their voices heard.

Over-Reliance on Existing Biosafety Measures and Performance

Just as we can question and debate the decision-making processes that led to the work and actions described in the previous examples in this chapter, we may also question and debate the oft-cited confidence and trust in biosafety processes that are frequently named as justifications for the safety of contentious work or as defences for the public.

In placing such trust in biosafety we need to remember that a biosafety policy is only effective if it is:

- A suitable and appropriate, well-informed and up to date policy,
- Is applied by well-trained people who understand the theories underpinning it,
- Is applied fully at all times,
- Is reviewed regularly or at specific non-regular times,
- Is subject to performance management processes, further training and *any other required changes* as necessary.

Can you say, hand on heart, that your biosafety procedures — and staff/student performance — can match up to this? How many events have been recorded in your biosafety logs this year? How do you know you are seeing all the events that have taken place?

I have worked overseas and with scientists and science educators from many countries — most of them in the 'developing countries' group. These have included former Soviet states (Ukraine, Georgia, Tajikistan, Uzbekhistan), North African states (Libya, Tunisia, Morocco, Egypt, Algeria), the Middle East (Jordan, Iraq, Pakistan) and the Far East (Japan and Indonesia). I have also worked with scientists and science educators in the US, Canada, Australia, and obviously the UK. Some of the tales I have heard from my colleagues in some of these countries are truly awful. Here are a few:

- A pregnant researcher drinks sodium hydroxide from a water bottle in the lab fridge — in an unlabelled drinking water bottle amongst *real* bottles of drinking water (a miscarriage resulted),

- Burns to hands and face due to the use of wrongly labelled agents at the bench,
- Staff becoming infected with pathogens from samples stored in the same fridge as the lunches,
- Inadequate procedures in storing pathogens resulting in cross-infection,
- Use of domestic freezers in labs because professional freezing equipment cannot be afforded,
- Staff failing to recognise or understand basic biosafety and biohazard signs and symbols,
- Live agents being released into the external drains or burned outside the lab in the open air,
- Live agents being taken home by staff,
- Personal belongings kept on the bench,
- Staff refusing to wear appropriate personal protective equipment,
- Faulty autoclaves failing to sterilise as required — unrecognised for weeks,
- Staff being contaminated in the eyes with human tissue samples but failing to get checked out, or to review systems,
- Incinerators breaking down regularly so waste put into the general rubbish instead,
- Staff failing to wash hands according to procedures,
- Centrifuges and vortex mixers 'fixed' with blu-tak and bits of pencil,
- A newly-built BSL3 lab (developing country) with the showers in the wrong place and regular failures in directional air flow,
- Samples kept in domestic freezers at home for lack of appropriate equipment,
- Students refusing to adopt 'normal' biosafety procedures because 'it is their democratic right to do so' (newly 'democratic' country),
- And so on.....

Need I point out that some of these also occurred in labs in 'developed' countries! Have you ever looked at the 'overlyhonest-methods' feed on Twitter? While much of the content is jokey, I have spoken to science students who say that it contains more than a little 'actuality'.

Let us look now at some evidence around the 'strength' of existing biosafety policies and processes in the US and the UK. See if you retain your faith in them after reading the next sections.

Various UK Biosafety and Biosecurity Breaches

In December 2014, Ian Sample, the science editor of *The Guardian* (a UK broadsheet), published an article entitled *Revealed: 100 safety breaches at UK labs handling potentially deadly disease.*[23] Sample is a scientist as well as a journalist, with a PhD in Biomedical Materials from Queen Mary's, University of London. Noting that the UK has nine CL4 labs (containment level 4), he reviewed reports obtained by the newspaper from the UK Health and Safety Executive (HSE) covering the previous five years. His article recounted some sobering evidence of the failings of biosafety processes around the country:

- More than 70 incidents at government, university and hospital labs were serious enough to be investigated,
- Some were serious enough to warrant legal action as a response to the events and others triggered enforcement letters or led to prohibition and crown notices (orders which suspend or stop work),
- UK labs handling the most dangerous pathogens had reported more than 100 accidents or near-misses in the past five years (it is worth pointing out here that reporting mistakes and mishaps is a good thing - and we should not censure people who admit to their mistakes),
- In one case, live anthrax had been sent from a government facility to other labs in the UK because tubes of live and heat-inactivated anthrax samples got mixed up
  - o at one receiving site, the anthrax was handled in a higher containment lab, meaning those staff were safe,

---

[23] Sample, I. (2014) Revealed: 100 safety breaches at UK labs handling potentially deadly diseases. *The Guardian* 4/12/14. Available on: http://www.theguardian.com/science/2014/dec/04/-sp-100-safety-breaches-uk-labs-potentially-deadly-diseases. Accessed 21/12/15. Summary in text courtesy of Guardian News & Media Ltd.

- o in another lab, the samples were never opened,
- o but in another site, scientists opened the tubes in a less secure lab and got to work on the open bench, resulting in exposure.

- Another case involved the failure of an air-handling system at a large animal lab working on FMD,
- Tears were found in isolation suits at a government lab handling the Ebola virus (Porton Down),
- The reports compiled by the HSE describe at least 116 incidents and 75 completed investigations since April 2010 at laboratories where the most dangerous organisms are handled,
- All of the investigations were prompted by reports from lab managers who are obliged by law to tell the HSE when an accident or near-miss happens at their facility.
  (Summary of the article courtesy of Guardian News & Media Ltd).

There was more, but you will agree, I hope, that this is problematic. Bear in mind that we cannot criticise people for reporting problems. I would rather see a lot of reports than experience an outbreak because people wished to protect their reputations by hiding mistakes. Sample and an expert who reviewed the reports for *The Guardian* stated:

> The figures amount to one investigation every three weeks at secure laboratories that are designed to carry out research on pathogens that can cause serious illness and spread into the community. Some of the organisms are lethal and have no vaccines or treatments........ the reports for the Guardian.... taken together..... revealed failures in procedures, infrastructure, training and safety culture at some British labs.

Many of the incidents were one-off, almost inevitable human mistakes, such as spillages of infectious pathogens. Others were down to old equipment and safety clothing. The most serious accidents arose from chains of mistakes that happened one after the other, and were often only discovered later.

One problem here was that in sending out live samples, the staff from the original labs were safe because they followed the right procedures to protect themselves (although if they could mix up live and inactivated anthrax in tubes, one wonders for how long), whereas those at the receiving end may not have been competent to protect themselves — having only expected inactivated anthrax. This is 'human error' writ large, with potentially devastating consequences.

The same lab had apparently been involved in major issues previously, including the shipping of 700 samples of *mycobacterium bovis* (causes TB in cattle but can also affect humans badly) to another lab without the managers knowing the organisms were still viable. The wrong equipment had been allocated for destruction of the organism and staff were not trained in the right procedure, even when staff raised concerns.

Sample's article listed 'Poor management, inadequate training, inappropriate procedures, equipment failures, human error and plain bad luck' as all playing a part in the incidents cited. But more worryingly, it was those incidents arising from a cascade or chain of lesser events that resulted in one or more large incidents that may give us more pause for thought. These lesser events, often ignored or not even noticed, can lead to massive outcomes through unwittingly 'lighting the blue touch paper.' Can we all say honestly that we have not been involved in many 'little' mistakes? Professor Brian Spratt, an infectious disease specialist at Imperial College, London was quoted in the article as saying 'What strikes me is that accidents do happen even in the best facilities, often due to operator error, or unrecognised breakdowns in containment measures.'

University of Birmingham, UK Smallpox Incident, 1978

A report by an expert[24] in 2009 had highlighted the risks of laboratories worldwide retaining their samples of the smallpox virus. While the WHO adopted Resolution 33.4 in 1980, advising all countries with stocks of the virus to destroy them, it had no powers to verify this if and when it was carried out. In 1975 over 70 labs worldwide

---

[24] Tucker, J. (2009) The Smallpox Destruction Debate: Could a Grand Bargain Settle the Issue? *Arms Control Today* website. Available on: https://www.armscontrol.org/act/2009_03/tucker. Accessed 30/12/15.

admitted to possessing stocks of the virus. An incident at the University of Birmingham, UK in 1978 resulted in the death of a medical photographer and the suicide of the laboratory head. The Birmingham incident was the subject of a government inquiry and report.[25] Although the source of the infection was identified (the lab below the room in which the photographer worked), the route of transmission was never fully established even when the University was prosecuted.

The facility at Birmingham had been turned down as a collaborating centre in future smallpox research by the WHO in 1977, prior to this incident occurring. It had, however, successfully passed inspections by the WHO in May 1978 and by the UK's Dangerous Pathogens Advisory Group or DPAG (now the Advisory Committee on Dangerous Pathogens, part of the HSE) in 1976. Due to the failure to acquire a role under the WHO future smallpox research in 1977, all work on the virus was expected to finish at Birmingham in late 1978 (a WHO approved decision). The Shooter report[26] later expressed concern about the fact that a facility that had been 'approved' by these and other authorities was, nevertheless, the site of a variola escape with a resulting infection.

Part of the investigation looked at how the facility had been able to continue with smallpox work despite not meeting either new UK standards developed following an outbreak at the London School of Hygiene and Tropical Medicine in 1973, or the WHO's new standards in 1978. Reasons given at the time for this included the fact that the Birmingham lab was a regional reference centre covering an area from which many people regularly travelled to tropical regions; the head of the lab, Professor Henry Bedson, was an experienced and reputable virologist with a reputation for meticulous work and safety standards; only a small number of staff worked on the virus and only

---

[25] Shooter, R.A. (1978) *Report On The Investigation Into The Cause Of The 1978 Birmingham Smallpox Occurrence*. Available on: https://www.gov.uk/government/uploads/system/uploads/attachment_data/file/228654/0668.pdf.pdf. Accessed 30/12/15.

[26] *Ibid.*

under his supervision and the staff had been subject to an up to date vaccination programme. However, the facility lacked several key elements of infrastructure, which with hindsight should probably have been used as justification to stop work there at an earlier date.

Janet Parker, the medical photographer, became the last person known to die of smallpox, in September 1978. Professor Bedson, the facility head, committed suicide apparently as a result of his perception of having failed to prevent the incident. The Shooter Report found that a chain of occurrences led to the infection of Parker (her parents were also quarantined). These included a promotion for Professor Bedson which had taken him away from the lab for greater periods of time; the delegation of work to a junior individual who was allowed to work unsupervised (a PhD student); the creeping in of a range of poor biosafety practices, and failure to use equipment properly. It was recognised that work at the lab had *increased* in 1978 as the deadline for closing the smallpox work approached — leading to greater work pressures and poor decisions being made to accommodate this. The Dangerous Pathogens Advisory Group's inspection and approval of the facility also came in for criticism, including that it had not carried out sufficient investigations and had misunderstood its remit.

These problems illustrate some of the typical issues that we have looked at already:

- Poor or insufficient staff training;
- Poor or lack of appropriate supervision;
- Problems with infrastructure (buildings, equipment);
- Problems with maintenance of these;
- Failure to recognise that changes in circumstances lead to the need for safety review.

More worryingly on the large scale, the Birmingham facility was approved to continue with its smallpox work by at least two major oversight bodies (DPAG and the WHO) *who thought they were acting in good faith and appropriately at the time.* How many times might similar inspections and decisions be allowing unsafe work now? Granted, hindsight is a wonderful thing, but surely we need to learn from experiences such as this (and others) that what seems right today may be revised in the face of just one incident. Should we not err on the side of caution as a matter of course? If this means upsetting *some* scientists *some* of the time, is it a price worth paying to protect the public and other staff?

Openness, not punishment

This is not just a problem of science or scientists. It is also a problem of management. Managers also need to take responsibility for all of these issues. Science staff may make mistakes and we will never get away from 'human error' — but once recognised or reported, actions need to be taken by managers to review procedures, identify failings, and revise activities to prevent recurrences. Blaming scientists will not help. It is only when scientists feel confident in bringing forward their concerns and, crucially, confident that appropriate and effective steps will be taken in response, that they will be able to, or want to, do so freely. If juniors observe seniors in 'covering up' or ignoring problems, they will simply learn that this is the way to go, and the problems will be perpetuated and embedded in practice. Punishment is not the answer — open communication of problems, followed by investigation, review and retraining is the best way forward. Plus — the sharing of mistakes and the lessons learned between peers and labs. Are we big enough to do this?

Various US Biosafety Breaches

A similar pattern of biosafety problems exists in the United States. The US Centers for Disease Control and Prevention (CDC) has had numerous problems in recent years. Given that the CDC labs are well-funded and carry out national and international functions, I will use them here as an example of what can happen even in such top-level facilities.

Papers obtained by a US newspaper in 2012[27] found that the airflow system in the building housing the Emerging Infectious Diseases Laboratory at the CDC facility in Atlanta, Georgia was not working correctly and appeared to have a leak into a hallway from a lab. This had been detected by a CDC safety inspector and appeared to involve a problem with the negative pressure system that is required in high-containment labs. The report also recounted how loss of electrical power had caused systems to fail in 2007 and how a door in another lab had been found *sealed with duct tape* in 2008.

Work at some labs and some shipments of materials were halted at the CDC in 2014 following a number of incidents involving federal labs. These included shipping live anthrax to other facilities, finding forgotten samples of live smallpox in storage at a federal lab and the shipping of a dangerous influenza sample to another facility.

The forgotten smallpox samples came to light during a clean-up at the Food and Drug Administration laboratory at the National Institutes of Health facility in Bethesda, Maryland.[28] This occurred despite smallpox being eradicated globally around 1980 and the only two legal locations for storage of the agent being at the CDC, Atlanta and in Novosibirsk, Russia. These samples were tested and found positive for the live smallpox virus, contradicting the opinions of some experts who did not think the samples would be viable after so long. The samples were all destroyed.

In June 2014 up to 75 CDC workers were potentially exposed to live anthrax samples in an incident involving the relocation of 'dead' samples from a BSL3 lab to a BSL2 lab on the same site. In this case, the transfer of the samples, which had failed to become inactivated following usual procedures to achieve this, were received by the lower level lab on 6 June without any written evidence or protocol in place

---

[27] Moisse, K. (2012) *Air Leak Sparks Safety Fears at CDC Bioterror Lab*, ABC News report. Available on: http://abcnews.go.com/Health/Wellness/cdc-atack-air-containment-problems/story?id=16557248. Accessed 21/12/15.
[28] Kaiser, J. (2014) Six vials of smallpox discovered in U.S. lab, Science News. Available on: http://news.sciencemag.org/health/2014/07/six-vials-smallpox-discovered-u-s-lab. Accessed 30/12/15.

to assure the staff that the samples were in fact inactivated. On 13 June, a scientist in the BSL 3 lab noticed unexpected growth on one of the anthrax sterility plates that had been used in the earlier work. This alerted staff to the problem. A report[29] outlined a chain of events and circumstances that led to the opportunity for this to happen.

CDC had also been involved in previous biosecurity incidents that were also noted in the 2014 report. In 2006, the Bioterrorism Rapid Response and Advanced Technology (BRRAT) laboratory in Atlanta, Georgia, had inadvertently transferred samples of live anthrax DNA to another national lab and to a private lab. On receipt, the national lab tested the samples and found viable *B.anthracis* present. As a result of this incident, CDC developed a new policy on shipping DNA materials of select agents (agents deemed dangerous to national security with access, handling and storage controlled by legislation). The 2014 anthrax incident did not need to follow this policy however as the shipment did not involve the preparation of DNA for transfer. Also in 2006, live *Clostridium botulinum* was shipped to another CDC facility having gone through ineffective inactivation processes. In 2009, an even more worrying incident involved the shipping of samples of *Brucella* to Laboratory Response Network labs *which had been going on since 2001*. It had been thought that the samples shipped were all of an attenuated vaccine strain, but on testing in 2009 it became apparent that this was not the case — the shipped strain *was actually a select agent* (subject to tight legal controls).[30]

Also in 2014, a culture of low-pathogenic avian influenza was, unrecognised at the time, cross-contaminated at a CDC facility with a highly pathogenic H5N1 strain of influenza and shipped to a government-run BSL-3, select agent laboratory. At the receiving lab, the culture was propagated and injected into chickens to observe them for signs of expected illness (relating to the disease sample scientists thought they had received). The chickens unexpectedly died, leading the scientists involved to test the culture they had received from

---

[29] Centers for Disease Control and Prevention (2014) *Report on the Potential Exposure to Anthrax.* Available on: http://www.cdc.gov/about/pdf/lab-safety/Final_Anthrax_Report.pdf. Accessed 30/12/15.
[30] *Ibid*, p. 3.

CDC. It turned out to be contaminated with the H5N1 virus. On investigation at the original CDC lab, the route of contamination was uncertain, but the report found a number possible ways in which it could have happened, all of which involved poor handling and poor biosafety practices.[31]

A CDC report[32] published in March 2015 revealed that two primates at the Tulane National Primate Research Center in Covington, Louisiana had been found to be ill with Melioidosis (Whitmore's Disease), a bacterial condition not found in North America. The investigation found that the agent involved, *Burkholderia pseudomallei*, was identical to one used at the Tulane centre for research. While the actual route of transmission was not identified, a range of options were highlighted as possible answers. These included problems with the correct use of protective clothing to prevent contamination of inner clothing allowing bacteria to be carried between labs. Inspectors noted that Tulane primate centre staff often went into the select agent lab without proper protective clothing. Select agent research was suspended at Tulane until certain remedial actions were taken, including training policy and practice in the use of protective clothing and equipment and improved entry and exit procedures to the animal enclosures.

A report by the National Research Council[33] in 2011 states that between 2003 and 2009, US government labs recorded almost

---

[31] Centers for Disease Control and Prevention (2014) *Report on the Inadvertent Cross-Contamination and Shipment of a Laboratory Specimen with Influenza Virus H5N1*, p. 5–6. Available on: http://www.cdc.gov/about/pdf/lab-safety/investigationcdch5n1contaminationeventaugust15.pdf. Accessed 30/12/15.

[32] Centers for Disease Control and Prevention (2015) Media Statement *Conclusion of Select Agent Inquiry into Burkholderia pseudomallei Release at Tulane National Primate Research Center*. Available on: http://www.cdc.gov/media/releases/2015/s0313-burkholderia-pseudomallei.html. Accessed 30/12/15.

[33] Committee on Risk Assessment of the Medical Countermeasures Test and Evaluation (MCMT&E) Facility at Fort Detrick, Maryland; National Research Council (2011) *Review of Risk Assessment Work Plan for the Medical Countermeasures Test and Evaluation Facility at Fort Detrick: A Letter Report*, p. 5. Available on: http://www.nap.edu/catalog/13265/review-of-risk-assessment-work-plan-for-the-medical-countermeasures-test-and-evaluation-facility-at-fort-detrick. Accessed 21/12/15.

400 incidents involving the potential release of select agents. These included:

- animal bites and scratches, 11 cases,
- needle stick or sharps injuries, 46,
- equipment mechanical failure, 23,
- personal protective equipment failure, 12,
- loss of containment, 196,
- procedural issues, 30.

A report by the Center for Infectious Disease Research and Policy (CIDRAP) at the University of Minnesota, stated that a CDC spokesman said, in response to an enquiry from CIDRAP, that select-agent labs are not identified for security reasons. This is an interesting issue in itself, as the general public is, in effect, kept 'out of the loop' in relation to the possible risks inherent in having labs in their area working on highly dangerous pathogens. While public confidence is important, there is an argument to be had about the balance between the public's right to know if they are living next door to a select agent lab, with neighbours regularly coming into contact with dangerous diseases (even if under containment) and the need for countries to maintain an active programme of research into medical aspects of these agents. I don't have an answer to this, but I suspect that it is an issue that will gain more attention in the near future as people want to know more in order to make informed decisions for themselves.

Lastly in this list of US examples, it was reported, and admitted, in 2015 that the Pentagon had inadvertently shipped live anthrax spores to 88 labs who shared it with 106 others, amounting to a total of 194 labs (at the last count reported on 1 September 2015) in 50 US states and across nine other countries.[34] The shipments originated at the US Army's Dugway Proving Grounds in Utah. An early estimate on 1 June 2015 said that live anthrax had gone to 24 labs in 11 states and

---

[34] Sisk, R. (2015) *Pentagon Now Says Army Mistakenly Sent Live Anthrax to All 50 States.* Available on: http://www.military.com/daily-news/2015/09/01/pentagon-says-army-mistakenly-sent-live-anthrax-all-50-states.html. Accessed 21/12/15.

to two other countries. Part of the problem here was the length of time it took for the number of lab contacts to be traced. Anything could have happened in this period — theft of materials at the very least. The anthrax strain involved was the Ames strain, the same as was used in the 2001 anthrax letters in the US. This is a virulent strain that caused the deaths of five people and infected 17 more in the letter attacks. The Pentagon claimed that the live spores did not constitute a risk to the public. Given that the receiving labs did not think they were receiving live anthrax, can this be a legitimate claim?

Summary

Any non-scientist, and many scientists, reading these accounts would be astonished that such events can occur in such numbers. How can air flow systems not work properly and this not be noticed in good time? How can drains be allowed to get into such a poor state that pathogens can escape from them into the environment? How can failing systems allow live pathogens to escape inactivation and this not be noticed? How can vials of live agent be mixed up with inactivated agents and be shipped in error? Why was it ever possible to accidentally send out live agents from labs when they were supposed to have been inactivated? Why were staff allowed to access a select agent lab without wearing appropriate protective clothing? How can stocks of live smallpox remain in storage for decades without anyone knowing? How can poorly trained or untrained staff be allowed to operate equipment and deal with dangerous pathogens?

It is unlikely that all lapses have been reported in either the UK or the US — and these are just two countries in which science is 'well regulated.' What of countries with less effective oversight, knowledge, training and suitable facilities? In my experience, labs in developing countries are usually operating at the level of education, training, knowledge, and motivation of their directors. Even if directors are keen to act appropriately and implement suitable policies to support biosafety and biosecurity, they themselves are often subject to pressures from managers that work against them achieving these ends. Biotechnology is a mushrooming industry around the world — how safe do we think a lot of facilities are, when many countries view them as a way of boosting national prosperity? If financial gain is the main aim, where does that put priority on being open about accidents and other failings?

In the West, we educate and train thousands of scientists from developing countries each year. It is incumbent on us to include not only stringent biosafety and bio-chemical security training in the education of such students (as well as our 'home' students), but also to emphasise the need to fully understand the theories underpinning these. This involves, of course, us providing a proper role model for all of our students by applying such theories and actions in our own practice. If we cannot be relied on to 'get it right,' what hope is there for juniors who qualify and go back to their countries as experts, with the view that cutting corners is acceptable?

It is against this background that scientists in labs in Europe and North America are telling us that it is safe to carry out challenging work on gain of function experiments — and others — because existing biosafety policies and processes are sound. It is clear from the foregoing examples that we cannot rely on the 'strength' of biosafety procedures to act also as bio-chemical security procedures. We need to address bio-chemical security directly and not view it as an 'add on' to biosafety. We need to bring the ethics of science practice into the debate here — The Ethics Toolkit questions outlined in earlier chapters allow us to assess our own practice for weak spots — and indicate when, where and how (to some extent) we can address these weaknesses in order to strengthen our resilience against bio-chemical security lapses. At least this is a start on a long and challenging journey to full bio-chemical security status.

An Interesting Exercise

Choose your favourite scenario of those illustrated above and try breaking down the progress of the scientific and science-related activities described in the scenario into the standard research stages. To each stage, try applying the ethical principles by considering the questions under each principle, as outlined in Chapters 5 and 6. You may find that had some or all of the principles described in earlier chapters been applied in the case of these scenarios, the outcomes may have been different.

Another way to practise using The Ethics Toolkit questions is to review some of your own work that had unintended consequences that you would have liked to avoid, or to have handled more effectively. Try looking back to the stages of your work and apply the ethical principles retrospectively. You may well be able to identify the exact stage of your work at which the ethical principle(s) was compromised — and by identifying the time, you can then see what you could have done to change the outcome.

Finally in this chapter, what do you now think about our apparent reliance on 'robust' biosafety procedures? Have these examples raised any concerns with you? Are you still sure that biosafety can 'cover' biosecurity?

Chapter 9

Closing Thoughts

Ethics is a subject that constantly surprises people. As I mentioned earlier, most people tend to confuse ethics with their personal values. This leads to the difficult situation in which they genuinely feel that they do not need to 'learn' ethics. Any ethics tutor reading this will recognise the need for a thick skin and a crash helmet when entering the Ethics Lesson arena. One of my colleagues, with his hand on the classroom door handle, once said to me in a worried tone: 'I am about to go over the top,' as if he were leaving the security of a First World War trench to step into enemy gunfire. I know the feeling.

Before I was privileged to learn about ethics, particularly research ethics, I only had a vague idea of what ethics *is*, like everyone else. I knew that it meant 'doing the right thing' and that 'being ethical' was something to be aspired to. But I had little or no idea about what it meant to be ethical in practice, other than holding to my own personal values. I had only vaguely contemplated the idea that we can each have personal values in private but use another set of values in public. Once this was brought to my attention (initially as an anthropologist), it made me re-evaluate a lot of my thoughts, ideals, and values.

The various issues that I have raised in previous chapters will assist you in evaluating your own work now that you have had a chance to think about them. While I started the book with reference to our responsibilities under the BTWC and the CWC, I hope that I have helped you to recognise the need for applied ethical practice for its own sake, as well as being a means through which you can meet your legal responsibilities under these two conventions.

Of course, you can also use The Ethics Toolkit to evaluate the work of your students, your colleagues and your professional rivals (there are bound to be some). Many of my students have come back to me years later and told how, after being introduced to ethics in specific classes, they were able to re-evaluate their own previous work, occasionally with frightening results. This is not something to be feared — look at it as a learning experience. Once you have spotted how to avoid a particular problem based on a real-life scenario, you will remember it for ever and are not likely to repeat the mistake. If you can recount your own 'bad' situations in class and over the water cooler by making a learning experience of them as an amusing (or terrifying) story about your own failures, so much the better. Some of my best sharing and teaching experiences with colleagues and students have involved telling them all about the various research 'messes' I have been involved in (not in dual use, I hasten to add) and encouraging them to spill their own versions — of which there have been plenty.

Another way of practising with The Toolkit questions is to get hold of your favourite academic paper and apply the questions to it. You may be surprised at what you find. Not all of the answers will be included in the paper, of course, as ethical behaviour is largely unseen, but at least you will be able to spot *where* the ethical questions ought to have been asked and *which* ethical principles were involved (and you may get the opportunity to ask questions in person if you know the author — assuming he or she does not take offence). In terms of passing on ethical expertise and competence, I find that using real academic papers in this way is a very useful exercise for learners — at any stage of their careers.

Do not be daunted by bio-chemical security ethics. It is simply another form of social responsibility. Effective engagement with ethics is largely a matter of common sense, courtesy, a grasp of the notion of protection of and care for others, thinking 'sideways' about your own research and that of others, following guidelines or regulations and applying experience.

Bear in mind that raising awareness of dual use risks and providing your colleagues and students with a skill set that allows them to identify potential dual use risks, is a waste of time if you do not then follow up with the implementation of appropriate activities to minimise the risks. The question then is, do the minimisation activities cause further problems in themselves? And does this mean you should not implement risk-minimising activities? The answer, of course, is no. But we all need some trial and error before we get things right.

Remember to start with ethics on 'Day One' of every project if you can. Consider ethical issues at all stages in the research process. Be prepared to make changes if necessary. Keep on reviewing the ethical situation as you progress through the research. We don't just 'do' ethics at the beginning and then get on with the 'real' work. Ethics pervades everything that we do, just like a pattern woven into a piece of fabric. As scientists we need to be always vigilant, aware that our benignly intended work may in fact have some unforeseen negative aspects that we need to minimise if possible.

Always consider how you would feel about misuse of the research if you were the research participant, the colleague, the end-user or an inadvertently-exposed person. What would you expect the original scientist to have reasonably done to protect you from harm?

Do not hesitate to ask for advice from colleagues, members of ethics panels, your Research Ethics Committee or professional association. Remember as well that you can share your own ethics expertise and experience with others, including ethics committees, as you become more familiar with The Ethics Toolkit and any other ethics frameworks you choose to use. Overall, learn from both your own and others' experiences and incorporate this into your own practice wherever possible. Aim to incorporate ethical awareness and application into your own teaching and sharing through courses, materials or discussion sessions. Only by building an effective level of applied ethics competency in the life sciences will we be able to effectively manage the many challenges posed to science by the needs of security. Sharing your experience and skills in ethics will enable you to contribute to the science community's ethics competency and encourage others to do the same.

Finally, good luck with your work. May it be always ethical.

The Ethics Toolkit in Brief

Section 1 The Rights of Others

The Autonomy Principle

The Ethics Toolkit: Implementing voluntary participation in science

- Do all my would-be participants or research/work colleagues know that they do not have to take part in my research if it has dual use potential?
- How can I assure myself that this is the case?
- Do all participants or research/work colleagues understand what the research/work is about and its potential implications?
- What do I need to do about this?
- Are any benefits or incentives offered? If so, are they agreed (by whom?) as appropriate?
- Am I exerting undue pressure on would-be participants or research/work colleagues?
- Is there any power relationship here? (Usually!)
- If so, how do I minimise or remove the potential effects of this? (Err…)
- Do those giving consent on behalf of others understand the implications of participation?
- Is the capacity of third party consent-givers acceptable and is their consent-giving relationship to the participant recognised as legitimate? (for example, are you assuming that all of your support staff and so on are 'happy' with what you are working on? How would you know that?)
- If I am undertaking some covert or indirect research, have I sought and found suitable advice and guidance from more experienced colleagues if necessary? (This is a major issue in potential dual use contexts — and should be **avoided** in civilian labs)
- Do I need to gain formal approval for my research from an ethics panel or review board? (How do you know they really understand the issues?)

- Is my research governed by any external body that needs to be consulted? (If in doubt, consult — it's too late when the pathogen is breeding happily among the local bird/animal/human population.)
- *Am* I seeking consent from reliable and valid would-be participants or research colleagues?
- *Should* I be seeking such consent?

The Ethics Toolkit: Implementing consent in science

- Do I need to obtain consent? [From whom? For what? When? How often?]
- Do I need to obtain written or simply oral consent?
- How do I record this?
- Does the participant/research/work colleague get a copy of the signed consent?
- What information do I need to give out in the consent process?
- Am I assuming consent from participants/research/work colleagues when it should be formally addressed?
- Am I satisfied that previously-obtained consent was obtained appropriately?
- Have my participants/research/work colleagues been given information about how they can leave the study/work/project?
- How often, and when, will I revisit consent with my participants/ research/work colleagues?
- Where and how will I store consent records?
- Have I considered all the issues that I can cover in the consent process?

The Privacy Principle

The Ethics Toolkit: Implementing privacy/security in science

- Do I need to publically identify participants/research/work colleagues?
- Do I need to offer any privacy to participants/research/work colleagues?

- How can I agree this with them?
- Do research/work colleagues need anonymity or confidentiality in terms of being identified with the research/work?
- How can I agree this with them?
- Are there any known risks or dangers for my colleagues in being associated with the work of our facility?
- Do I need to do anything to mitigate these risks or dangers through privacy mechanisms?
- What are the social and/or economic implications for colleagues in such cases?
- Can I actually manage and deliver what I am offering?
- Who has access to what, and where and when?
- Am I satisfied that there are sufficient and appropriate control measures in place?
- Can I control the security of my privacy measures?
- How can I securely manage the transfer of data from the field to the record store, or from lab to lab?
- Have I considered the possibility of re-use of the data in the future and how this may impact on anonymity and retention of key lists?
- And so on...

Section 2 Your Responsibilities in Science

The No Harm Principle

The Ethics Toolkit: Implementing 'do no harm' in science

- In the context of my work, what is harm?
- How would I recognise that?
- Who is it harmful to?
- Is there any reasonably foreseeable risk associated with participation in, or association with, my research/work?
- Who is affected?
- Immediately?
- Soon? (Can you define 'soon'?)
- Later? (Can you define 'later'?)
- Where may the presence of harm become apparent? (You can't see infection until symptoms appear)

- Can I reasonably reduce the risk of this harm?
- Is there a less harmful way in which I can achieve the desired outcomes from this research question?
- Am I prepared to stop, delay, or otherwise discontinue my work if a certain threshold is reached?
- Who decides what that threshold is, or should be?
- How can I effectively communicate all of this to participants in my work? (Remember that a far wider pool of people 'participate' in your work than you think — your family are exposed to *you*, your colleagues and their families are exposed, all people who visit your facility are exposed, and their families and friends after them — where do the ripples end?)
- How much information about this should I give out and in what form?
- If harm emerges during the project, how can I and participants communicate about it and do something about it?
- What measures can I and my colleagues put in place to minimise risk?
- Can we do anything to provide support for participants or associates who may be affected by harm?
- What *should* we be doing to achieve this?
- Have I considered how to manage any clashes between the values and norms of my country or workplace and those of people from other communities?
- How can we *all* go about agreeing on definitions of harm, and the necessary steps to respond to harm when it occurs?
- Where will all this be codified?
- Who will be responsible for oversight of this?
- How often will it all be reviewed and revised?

The Beneficence Principle

The Ethics Toolkit: implementing beneficence in science

- Am I doing this work primarily for the benefit of others or for myself?
- How would I know this?
- How would others know it?

- What constitutes a benefit?
- To whom?
- What is the work for?
- Is the benefit real or just projected by me for some point in the future?
- How do we know that the benefit will ever be needed?
- Can we justify work now on the basis of a possible 'need' for that benefit in the future?
- Do I offer (wittingly or unwittingly) direct or indirect benefits to participants/research/work colleagues? (Apart from their pay and work benefits — see Chapter 5 — *Can I offer incentives to participate?*)
- Why am I doing this?
- Is this right?
- How may these 'benefits' be best managed (offered, limited, be appropriate)?
- Do they amount to coercion to participate or to continue with the work?
- Will the benefits actually lead to further problems to the recipients or to me?
- Are the benefits I offer colleagues actually meant to bring further benefit to me?
- Should I be offering certain benefits to colleagues in the interests of fairness?
- Have I been withholding benefits from colleagues or juniors in order to keep my place in the pecking order?
- Who resources benefits?
- Is there a conflict of interest? (There usually is if you look closely enough).
- Where do the power relationships sit in this?
- Am I unduly influenced in my opinions on right and wrong in science by the benefits that my position as a scientist confers on me?
- How would I know that?
- Has anyone ever challenged me on this? How did I respond?

- If, from now on, all science publications had to be anonymous, and all prizes were abolished, what effect might that have on me and my attitude to my work?
- How would I manage my expectations about what science can do for me if I live in a country where scientists have to do what the government tells them to? (And cease from doing certain work if the government tells them to?)

The Responsible Dissemination Principle

The Ethics Toolkit: implementing responsible dissemination of science

- Am I aware of any current, recent or future potential for the misuse of my research?
- If so, how may I minimise the risk of such misuse through consideration of my
- dissemination methods?
- Are there options to be explored about how I could disseminate my work in other ways than the traditional routes at this time?
- Who can advise me on alternative dissemination methods?
- If I choose to redact or partially publish, what will I *really* suffer?
- What *advantage* may I gain through only partial publication?
- What other arrangements could I put in place to supply details to other scientists after I partially publish my work?
- What advantages could this give me?
- What disadvantages could it give me?
- Who will potentially be affected by any dual use of my work? Should they be warned?
- What impact would full or partial publication of my work have at a different time or place?
- Should I consider holding back some of my findings from certain types of publication or from certain groups?
- How would I disseminate those withheld findings to 'safe' destinations?

The Scholarship and CPD Principle

The Ethics Toolkit: implementing scholarship and CPD in science

- How can I keep up to date with methodological developments in my field?
- What further training and development do I need to work more effectively?
- How will I know that I need to be brought up to date?
- How much of my time should I spend on continuing professional development (CPD)?
- Do I have any responsibility to maintain the CPD of my colleagues/staff?
- How can I make applied ethics a regular part of the learning of my juniors and colleagues through CPD sessions?
- How can I spread my/our good practice (ethics and science) to other colleagues?
- How can I/we pick up the good practice (ethics and science) of colleagues and bring it into our facility?
- How open am I to change my practice (ethics and science) if I see a better way?
- If I am introduced to a more ethical way to carry out my work, am I prepared to change my work accordingly?
- Am I prepared to talk to my colleagues to persuade them to change their work?
- What issues are preventing me from changing my practice?
- What issues are preventing my colleagues from changing their practice?
- What can we do about this?

Some Further Reading that May Interest You

The following readings are intended to supplement the various sources cited in the text. This list is not exhaustive but is intended to provide you with further useful information if you are interested. All URLs checked on 30/12/15.

Crowley. M. (2009) Dangerous Ambiguities: Regulation of Riot Control Agents and Incapacitants under the Chemical Weapons Convention, Bradford Non-lethal Weapons Research Project, University of Bradford, October 2009. Available on: http://www.brad.ac.uk/acad/nlw/publications/BNLWRP Dangerous1.pdf

Dando, M.R. (2004) *Biotechnology Weapons and Humanity II*, 22 September 2006, Board of Science and Education, London: British Medical Association.

Dando, M.R. and Rappert, B. (2005) *Codes of Conduct for the Life Sciences: Some Insights from UK Academia*, Bradford Briefing Paper No. 16 (Second Series). Available on: http://www.brad.ac.uk/acad/sbtwc/briefing/BP_16_2ndseries. pdf.

Fiddler, D. and Gostin, L.O (2008) *Biosecurity in the Global Age: Biological weapons, Public Health and the Rule of Law*, Stanford, California: Stanford University Press.

Hay, A.W.M., Smith, G.L., Smith, D., McCauley, J., Sture, J.F., Drew, T. and Ashcroft, R. (Dual Use Research of Concern Working Group) (2015) *Managing risks of misuse associated with grant funding activities* A joint Biotechnology and Biological Sciences Research Council (BBSRC), Medical Research Council (MRC) and Wellcome Trust policy statement. Available on: http://www. wellcome.ac.uk/About-us/Policy/Policy-and-position-statements/wtx026594. htm.

International Committee of the Red Cross (2002) *Biotechnology, weapons and humanity*. Available on: https://www.icrc.org/eng/resources/documents/misc/5hvjux.htm. Accessed 30/12/15.

InterAcademies Panel (IAP) (2005) *Statement on Biosecurity*, The InterAcademy Panel on International Issues. Available on: http://www.interacademies.net/ ?id=5405 (requires a log-in).

Italy (2005) *Codes of Conduct for Biological Scientists*, BWC/MSP/2005/MX/ WP.34, Meeting of Experts, Geneva. Available on: http://www.opbw.org/ new_process/mx2005/bwc_msp.2005_mx_wp34_E.pdf.

Japan (2005) *Codes of Conduct for Scientists: A View From Analysis of the Bioindustrial Sectors*, BWC/MSP/2005/MX/WP.22, Meeting of Experts, 13–24 June 2005, Geneva. Available on: http://www.opbw.org/new_process/mx2005/bwc_ msp.2005_mx_wp22_E.pdf.

Minehata. M. (2010) *An Investigation of Biosecurity Education for Life Scientists in the Asia-Pacific Region*, Research Report for the Wellcome Trust Project on 'Building a Sustainable Capacity in Dual-use Bioethics.'

Minehata, M. and Sture, J.F. (2010) Promoting Dual-Use Education for Life Scientists: Resources and Activities. Guest Editorial *Applied Biosafety* 15(4), 164.

Minehata, M., Sture, J., Shinomiya, N. and Whitby, S. (2011) Implementing Biosecurity Education: Approaches, Resources and Programmes. *Science and Engineering Ethics* 19(4), 1473–1486. Available on: http://link.springer.com/article/10.1007/s11948-011-9321-z.

Minehata, M. and Sture, JF. (2012) *National Series* First five editions of the Series: Pakistan, Ukraine, Georgia, Tajikistan and Azerbaijan. Available on: http://www.worldscientific.com/worldscibooks/10.1142/q0027#t=suppl

Minehata, M., Sture, J., Shinomiya, N., Whitby, S. and Dando, M. (2013) Promoting Education of Dual-Use Issues for Life Scientists: A Comprehensive Approach. *Journal of Disaster Research, Special Edition* 8(4), 674–685.

Novossiolova, T. and Sture, J. (2012) Towards the responsible conduct of Scientific Research: Is Ethics Education enough? Special Edition of *Medicine, Conflict and Survival* 28(1), 73–84. Available on: http://www.tandfonline.com/toc/fmcs20/28/1.

Pearson, G. (2002) *Relevant Scientific and Technological Developments for the First CWC Review Conference: The BTWC Review Conference Experience* First CWC Review Conference, Paper No. 1. Dept of Peace Studies, University of Bradford. Available on: http://www.brad.ac.uk/acad/scwc/cwcrcp/cwcrcp_1.pdf.

Revill. J. (2010) *Developing Metrics and Measures for Dual-Use Education,* Research Report for the Wellcome Trust Project on 'Building a Sustainable Capacity in Dual-use Bioethics.'

Revill. J (2009) *Biosecurity and Bioethics Education: A Case Study of the UK Context.* Research Report for the Wellcome Trust Project on 'Building a Sustainable Capacity in Dual-Use Bioethics.

Rappert, B. (ed) (2010) *Education and Ethics in the Life Sciences: Strengthening the Prohibition of Biological Weapons,* Canberra: Australian National University e-press. Available on: http://epress.anu.edu.au/education_ethics.html.

Rappert. B. (2009) *Experimental Secrets.* Landham, MD: University Press of America.

Selgelid, M.J. and Weir, L. (2010) Reflections on the synthetic production of poliovirus. *Bulletin of Atomic Scientists,* 66(3). (Requires a subscription log-in but also accessible by Open Athens log-in). Available on: http://thebulletin.org/2010/may/reflections-synthetic-production-poliovirus.

Sture, J.F. (2010) *Statement on behalf of the National Defense Medical College, Japan and Bradford Disarmament Research Centre, UK,* to the Meeting in December 2010 Of The States Parties To The Convention On The Prohibition Of The Development, Production And Stockpiling Of Bacteriological (Biological) And Toxin Weapons And On Their Destruction. Available on: http://www.unog.ch/ (Click on BTWC and Meetings 2010 — Meeting of States Parties).

Sture, J.F. and Minehata, M. (2011) *More BTWC Education Needed for Life Scientists* Op-Ed. Virtual Biosecurity Centre. Available on: http://virtualbiosecuritycenter. org/blog/op-ed-more-btwc-education-needed-for-life-scientists

Sture, J. and Whitby, S. (2012) Guest Editorship *Medicine, Conflict and Survival* 28(1). Special Issue: Preventing the Hostile Use of the Life Sciences and Biotechnologies: Fostering a Culture of Biosecurity and Dual Use Awareness. Available on: http://www.tandfonline.com/toc/fmcs20/28/1.

Sture, J., Minehata, M. and Shinomiya, N. (2012) Looking at the formulation of national biosecurity education action plans. Special Edition of *Medicine, Conflict and Survival* 28(1), 85–97. Available on: http://www.tandfonline.com/toc/fmcs20/28/1.

Sture, J. and Whitby, S. (2012) Conclusions. Special Edition of *Medicine, Conflict and Survival* 28(1):99–105. *Preventing the Hostile Use of the Life Sciences and Biotechnologies: Fostering a Culture of Biosecurity and Dual Use Awareness.* Available on: http://www.tandfonline.com/toc/fmcs20/28/1.

Sture, J., Whitby, S. and Perkins, D. (2013) Biosafety, Biosecurity and Internationally Mandated Regulatory Regimes: Compliance Mechanisms for education and global health security *Medicine, Conflict and Survival,* 29(4), 289–321. Available on: http://www.tandfonline.com/toc/fmcs20/29/4.

Sture, J. (2013) *International Biosecurity: Engagement between American and MENA Scientists* (commissioned consultancy paper providing suggestions for new, transformative bioengagement opportunities between United States and MENA countries) as part of a Project on Advanced Systems and Concepts for Countering Weapons of Mass Destruction (PASCC), Center on Contemporary Conflict, Naval Postgraduate School. Available on: http://www.worldscientific. com/worldscibooks/10.1142/q0027#t=suppl

Sture, J. (ed) (2013) *Yearbook of Biosecurity Education 2012* Bradford: University of Bradford. ISBN 978 1 85143 271 4. Available on: http://www.worldscientific. com/worldscibooks/10.1142/q0027#t=suppl

Sture, J. (2014) Dual Use. In *Global Encyclopedia of Global Bioethics* (Springer). Available on: http://www.springerreference.com/index/chapterdbid/ 398716.

There are thousand more sources you can look at in both the general and the science media. Try an internet search for 'biosecurity', 'dual use', 'biological weapons,' and 'chemical weapons' for starters.

Relevant Sections of the Biological and Toxin Weapons Convention for Scientists

http://www.opbw.org/convention/conv.html

Article I

Each State Party to this Convention undertakes never in any circumstances to develop, produce, stockpile or otherwise acquire or retain:

(1) Microbial or other biological agents, or toxins whatever their origin or method of production, of types and in quantities that have no justification for prophylactic, protective or other peaceful purposes;
(2) Weapons, equipment or means of delivery designed to use such agents or toxins for hostile purposes or in armed conflict.

Article III

Each State Party to this Convention undertakes not to transfer to any recipient whatsoever, directly or indirectly, and not in any way to assist, encourage, or induce any State, group of States or international organizations to manufacture or otherwise acquire any of the agents, toxins, weapons, equipment or means of delivery specified in article I of this Convention.

Article IV

Each State Party to this Convention shall, in accordance with its constitutional processes, take any necessary measures to prohibit and prevent the development, production, stockpiling, acquisition, or retention of the agents, toxins, weapons, equipment and means of delivery specified in article I of the Convention, within the territory of such State, under its jurisdiction or under its control anywhere.

Article V

The States Parties to this Convention undertake to consult one another and to cooperate in solving any problems which may arise in relation to the objective of, or in the application of the provisions of, the Convention. Consultation and Cooperation pursuant to this article may also be undertaken through appropriate international procedures within the framework of the United Nations and in accordance with its Charter.

Article VI

(1) Any State Party to this convention which finds that any other State Party is acting in breach of obligations deriving from the provisions of the Convention may lodge a complaint with the Security Council of the United Nations. Such a complaint should include all possible evidence confirming its validity, as well as a request for its consideration by the Security Council.

(2) Each State Party to this Convention undertakes to cooperate in carrying out any investigation which the Security Council may initiate, in accordance with the provisions of the Charter of the United Nations, on the basis of the complaint received by the Council. The Security Council shall inform the States Parties to the Convention of the results of the investigation.

Article IX

Each State Party to this Convention affirms the recognized objective of effective prohibition of chemical weapons and, to this end, undertakes to continue negotiations in good faith with a view to reach early agreement on effective measures for the prohibition of their development, production and stockpiling and for their destruction, and on appropriate measures concerning equipment and means of delivery specifically designed for the production or use of chemical agents for weapons purposes.

Article X

(1) The States Parties to this Convention undertake to facilitate, and have the right to participate in, the fullest possible exchange of equipment, materials and scientific and technological information for the use of bacteriological (biological) agents and toxins for peaceful purposes. Parties to the Convention in a position to do so shall also cooperate in contributing individually or together

with other States or international organizations to the further development and application of scientific discoveries in the field of bacteriology (biology) for prevention of disease, or for other peaceful purposes.

(2) This Convention shall be implemented in a manner designed to avoid hampering the economic or technological development of States Parties to the Convention or international cooperation in the field of peaceful bacteriological (biological) activities, including the international exchange of bacteriological (biological) and toxins and equipment for the processing, use or production of bacteriological (biological) agents and toxins for peaceful purposes in accordance with the provisions of the Convention.

Relevant sections of the Chemical Weapons Convention for Scientists

https://www.opcw.org/chemical-weapons-convention/download-the-cwc/

ARTICLE I

General obligations

1. Each State Party to this Convention undertakes never under any circumstances:
   (a) To develop, produce, otherwise acquire, stockpile or retain chemical weapons, or transfer, directly or indirectly, chemical weapons to anyone;
   (b) To use chemical weapons;
   (c) To engage in any military preparations to use chemical weapons;
   (d) To assist, encourage or induce, in any way, anyone to engage in any activity prohibited to a State Party under this Convention.

2. Each State Party undertakes to destroy chemical weapons it owns or possesses, or that are located in any place under its jurisdiction or control, in accordance with the provisions of this Convention.

3. Each State Party undertakes to destroy all chemical weapons it abandoned on the territory of another State Party, in accordance with the provisions of this Convention.

4. Each State Party undertakes to destroy any chemical weapons production facilities it owns or possesses, or that are located in any place under its jurisdiction or control, in accordance with the provisions of this Convention.

5. Each State Party undertakes not to use riot control agents as a method of warfare.

ARTICLE II
Definitions and criteria

For the purposes of this Convention:

1. "Chemical Weapons" means the following, together or separately:

   (a) Toxic chemicals and their precursors, except where intended for purposes not prohibited under this Convention, as long as the types and quantities are consistent with such purposes;

   (b) Munitions and devices, specifically designed to cause death or other harm through the toxic properties of those toxic chemicals specified in subparagraph (a), which would be released as a result of the employment of such munitions and devices;

   (c) Any equipment specifically designed for use directly in connection with the employment of munitions and devices specified in subparagraph (b).

2. "Toxic Chemical" means:

   Any chemical which through its chemical action on life processes can cause death, temporary incapacitation or permanent harm to humans or animals. This includes all such chemicals, regardless of their origin or of their method of production, and regardless of whether they are produced in facilities, in munitions or elsewhere. (For the purpose of implementing this Convention, toxic chemicals which have been identified for the application of verification measures are listed in Schedules contained in the Annex on Chemicals.)

3. "Precursor" means:

   Any chemical reactant which takes part at any stage in the production by whatever method of a toxic chemical. This includes any key component of a binary or multicomponent chemical system.

   (For the purpose of implementing this Convention, precursors which have been identified for the application of verification measures are listed in Schedules contained in the Annex on Chemicals.)

4. "Key Component of Binary or Multicomponent Chemical Systems" (hereinafter referred to as "key component") means:

   The precursor which plays the most important role in determining the toxic properties of the final product and reacts rapidly with other chemicals in the binary or multicomponent system.

5. "Old Chemical Weapons" means:

   (a) Chemical weapons which were produced before 1925; or

   (b) Chemical weapons produced in the period between 1925 and 1946 that have deteriorated to such extent that they can no longer be used as chemical weapons.

6. "Abandoned Chemical Weapons" means:

   Chemical weapons, including old chemical weapons, abandoned by a State after 1 January 1925 on the territory of another State without the consent of the latter.

7. "Riot Control Agent" means:

   Any chemical not listed in a Schedule, which can produce rapidly in humans sensory irritation or disabling physical effects which disappear within a short time following termination of exposure.

8. "Chemical Weapons Production Facility":

   (a) Means any equipment, as well as any building housing such equipment, that was designed, constructed or used at any time since 1 January 1946:

       (i) As part of the stage in the production of chemicals ("final technological stage") where the material flows would contain, when the equipment is in operation:

(1) Any chemical listed in Schedule 1 in the Annex on Chemicals; or

(2) Any other chemical that has no use, above 1 tonne per year on the territory of a State Party or in any other place under the jurisdiction or control of a State Party, for purposes not prohibited under this Convention, but can be used for chemical weapons purposes;

or

(ii) For filling chemical weapons, including, inter alia, the filling of chemicals listed in Schedule 1 into munitions, devices or bulk storage containers; the filling of chemicals into containers that form part of assembled binary munitions and devices or into chemical submunitions that form part of assembled unitary munitions and devices, and the loading of the containers and chemical submunitions into the respective munitions and devices;

(b) Does not mean:

(i) Any facility having a production capacity for synthesis of chemicals specified in subparagraph (a) (i) that is less than 1 tonne;

(ii) Any facility in which a chemical specified in subparagraph (a) (i) is or was produced as an unavoidable by-product of activities for purposes not prohibited under this Convention, provided that the chemical does not exceed 3% of the total product and that the facility is subject to declaration and inspection under the Annex on Implementation and Verification (hereinafter referred to as "Verification Annex"); or

(iii) The single small-scale facility for production of chemicals listed in Schedule 1 for purposes not prohibited under this Convention as referred to in Part VI of the Verification Annex.

9. "Purposes Not Prohibited Under this Convention" means:

(a) Industrial, agricultural, research, medical, pharmaceutical or other peaceful purposes;

(b) Protective purposes, namely those purposes directly related to protection against toxic chemicals and to protection against chemical weapons;

(c) Military purposes not connected with the use of chemical weapons and not dependent on the use of the toxic properties of chemicals as a method of warfare;

(d) Law enforcement including domestic riot control purposes.

10. "Production Capacity" means:

The annual quantitative potential for manufacturing a specific chemical based on the technological process actually used or, if the process is not yet operational, planned to be used at the relevant facility. It shall be deemed to be equal to the nameplate capacity or, if the nameplate capacity is not available, to the design capacity. The nameplate capacity is the product output under conditions optimised for maximum quantity for the production facility, as demonstrated by one or more test-runs. The design capacity is the corresponding theoretically calculated product output.

11. "Organisation" means the Organisation for the Prohibition of Chemical Weapons established pursuant to Article VIII of this Convention.

12. For the purposes of Article VI:

(a) "Production" of a chemical means its formation through chemical reaction;

(b) "Processing" of a chemical means a physical process, such as formulation, extraction, and purification, in which a chemical is not converted into another chemical;

(c) "Consumption" of a chemical means its conversion into another chemical via a chemical reaction.

You can read the remaining articles of the CWC if you need to, at: https://www.opcw.org/chemical-weapons-convention/download-the-cwc/.

Index

A

Academic and scientific freedoms, 25, 63, 66, 108, 183, 187, 189, 241, 245, 248
  and ethics, 107
  in H5N1 research, 247
  may come second to voters' concerns in eyes of governments when in difficult situations, 132
  nature and scope of, vi
  traditional assumptions no longer tenable, 7–8
  attacks on, vi
  going where the science takes us, 25
  need to rethink traditional freedoms, 3
  no longer absolute, 5
  traditional assumptions in, vii
Academic and Scientific Practice
  interventions by non-scientists, vii

Academic practice
  being prepared to modify practice, publication and communications occasionally, 4
  ethics as a regular part of, ix
  identifying and responding to biosecurity issues in, ix
  interventions by non-scientists, ix
  knowledge, experience and practice as hazards to traditional freedoms, see also Scientific practice, 9
  public scrutiny of, ix
  retaining influence over, ix
  scientists increasingly held to account, vii
  the ethics toolkit as one of a number of decision-making processes, 5
Accident logs, 14
Accountability for scientific practice and outcomes, 5, 7, 14, 39, 49, 60
  increasingly important, 248